Language and Representation

THE DEVELOPING BODY AND MIND

Series Editor: Professor George Butterworth, *Department of Psychology, University of Stirling.*

Designed for a broad readership in the English-speaking world, this major series represents the best of contemporary research and theory in the cognitive, social, abnormal and biological areas of development.

1. *Infancy and Epistemology: An Evaluation of Piaget's Theory*
   George Butterworth (ed.)
2. *Social Cognition: Studies on the Development of Understanding*
   George Butterworth and Paul Light (eds)
3. *The Mental and Social Life of Babies: How Parents Create Persons*
   Kenneth Kaye
4. *Evolution and Developmental Psychology*
   George Butterworth, Julie Rutkowska and Michael Scaife (eds)
5. *The Child's Point of View: The Development of Cognition and Language*
   M. V. Cox
6. *Developmental Psychology in the Soviet Union*
   Jaan Valsiner
7. *Language and Representation: A Socio-Naturalistic Approach to Human Development*
   Chris Sinha

Forthcoming:

The Developing Brain: Environmental Impact on Human Behaviour
T. C. Jordan and Frances Williamson

Child Prodigies
John Radford

An Introduction to the Psychology of Children's Drawings
G. V. Thomas and A. M. J. Silk

Language and Representation

A Socio-Naturalistic Approach to Human Development

CHRIS SINHA
Lecturer in Psychology
University of Utrecht

HARVESTER · WHEATSHEAF
NEW YORK LONDON TORONTO SYDNEY TOKYO

First published 1988 by
Harvester • Wheatsheaf
66 Wood Lane End, Hemel Hempstead
Hertfordshire HP2 4RG
A division of
Simon & Schuster International Group

Printed and bound in Great Britain by
Billing & Sons Ltd, Worcester

British Library Cataloguing in Publication Data

Sinha, Chris, *1951–*
Language and representation: a socio-naturalistic approach to human development.
—(The Developing body and mind; no. 7).
1. Language. Psychological aspects.
Theories
I. Title II. Series
401′.9
ISBN 0–7108–0626–4

1 2 3 4 5 92 91 90 89 88

To my family and friends

Contents

List of Figures

List of Tables

Foreword

It is indeed a pleasure to welcome this seminal book by Chris Sinha to the *Developing Body and Mind Series*. The aim of this Series has been to promote original thinking which encompasses biological, social and cultural influences on problems of human development. Of all the mysteries of development, the acquisition of language most fundamentally links human nature with human culture and it has proved a difficult theoretical task to develop a vocabulary that captures this essential link.

It has been Chris Sinha's achievement, in his novel approach to the relation between language and representation, to work out a social-materialist basis for development. He proposes a co-evolutionary view of the relationship between culture and biology, especially in the development of symbolization through communication. Rather than maintaining a dichotomy between nature and culture the developmental approach makes it obvious that, from the perspective of the infant who must acquire language, all that is accessible in culture exists in nature. Culture in its material aspect is accessible to the child because it embodies design features that are specifically related to the properties of the child's mind. In particular, he argues that infancy is the special biological mechanism through which cultural transmission occurs. The infant discovers and comes to represent the socially standard forms of material objects, such as tools and the canonical forms of speech. Interpersonal communication provides the material basis for speech, signs and symbols, through carefully structured opportunities for mind to engage with the spoken culture. But even in explaining Chris Sinha's enterprise in this way I have introduced a dichotomy which it is the purpose of the book to overcome. Although it is a complex undertaking, this book shows how we may progress in the aim of linking materialistic with social approaches to the development of mind.

George Butterworth
University of Stirling
March 1988

Preface

As the title of this book indicates, the problem of representation is at the heart of its inquiry into the developmental psychology of language. Representation, as Mandler (1984) notes, has a dual meaning in psychological theory. In the first place, it can refer to the signs or symbols which human beings produce and exchange. In the second place, it can refer to the knowledge and beliefs underlying sign-using and other behaviours, and the structure and organization of such knowledge and beliefs. Representation, then, is at once a semiotic and a cognitive category. Recent developments in cognitive science have thrust the concept of representation in its second sense, that of the structure of knowledge, to the forefront of contemporary psychology. This has been at the expense, however, of a sustained attempt to relate this concept of representation to traditional and modern approaches to the theory of the sign.

Within the computational 'modern synthesis' which dominates contemporary cognitive psychology, the problem of representation is viewed almost exclusively as a question of formalization. The nature of representation as signification, and its role in the social constitution of subjectivity, is deemed from this point of view irrelevant to scientific concerns. It is perhaps not surprising, then, that it has been within such discourses as psychoanalysis and literary criticism that semiotic (or semiological, according to one's preference) theory has recently flourished; still less, that where cognitive psychologists notice such developments at all, they typically view them with, at best, indifference, and, perhaps more frequently, outright suspicion.

The reason for the truncation in modern science of the notion of representation is to be sought, I shall suggest, in a set of related dualisms which have haunted and distorted the mainstream of Western thought since Descartes cast the problem of knowledge firmly in terms of the individual subject. These dualisms—between subject and object, individual and society, nature and culture, mind and reality—constitute both the power and the limiting conditions of that form of rationality which the technologically advanced, bureaucratic societies of both 'West' and 'East' embody. In the contemporary world, 'critical' theory subsists in the interstices of an objectivity which is the domain of a distributed and scientized authority, while our own subjectivities are dispersed and fragmented by machineries of power and desire realized through the commodity/sign. This book was conceived as an attempt to recover a critical dimension in the scientific study of the development of language and representation, while according due recognition to the important insights afforded by advances in cognitive science.

Within the (historically) largely positivist landscape of psychological science, developmental theory has long occupied the 'radical' ground, insisting against the once-dominant reductionist theories of behaviourism upon the autonomy of mind and mental processes. In this respect, Piaget's genetic epistemology has been seen as a forerunner of the 'mentalistic' theories now dominant throughout psychology (Boden, 1979), and it remains a central reference point for any investigation of the origins and development of representation.

More recently, however, developmental psychologists have become critical of many of Piaget's specific propositions, particularly those implying infantile 'egocentrism' and logical incompetence, and of his neglect of contextual and communicative dimensions of reasoning. As they became more aware of the importance of studying the social-ecological matrix within which human development naturally occurs (Bronfenbrenner, 1979; Cole *et al.*, 1971, 1978), and of the nature of the experimental setting itself as a 'socio-dialogic context' (Karmiloff-Smith, 1979), developmental psychologists began in the 1970s to forge a more 'social' approach to cognitive development and cognitive processes than that associated with classical Piagetian theory (Richards, 1974; Butterworth and Light, 1982; Richards and Light, 1986). In particular, a number of workers have recently sought to integrate Vygotsky's historical-cultural approach with current findings in the field of linguistic and pre-linguistic communicative development (Lock, 1978, 1980; Wertsch, 1985).

The study of language acquisition underwent a parallel and related shift, in the 1970s, from an initial preoccupation—inspired by Chomsky's transformational generative linguisitics—with syntax, to an increasing emphasis on semantics, and later pragmatics and discourse (Brown, 1973; Bruner, 1975; Greenfield, 1978; Karmiloff-Smith, 1979; Beveridge, 1982; Kuczaj, 1984). Recent developments in linguistic theory, in particular text-linguistic theory and neo-functional grammars (Beaugrande and Dressler, 1981; Givón, 1979; Dik, 1981; Dirven and Fried, 1987) testify also to the increasing importance attached by at least some workers within that discipline to supra-sentential and contextual factors in the formation of linguistic structure, and to the analysis of conversational discourse.

It is noteworthy too that 'developmental' or diachronic questions are re-emerging as problematic issues in linguistic theory (Bailey and Harris, 1985), and that many of the questions raised by developmental linguists regarding the traditional *langue–parole* distinction are similar to those raised by discourse and text linguists. The prevailing neglect within mainstream linguistics and cognitive psychology of the problem of development—or, more accurately, the formulation of developmental questions in terms of an inadequate opposition of learned versus innate mechanisms—is symptomatic of the closure of its concept of representation, not only to social and historical considerations, but also to evolutionary biological ones.

Developmental psychology—particularly Piaget's genetic epistemology—has historically been associated with unorthodox interpretations of evolutionary biological processes, and it is interesting that the current crisis affecting the neo-Darwinian synthesis is accompanied by a renewed interest on the part of some biologists in epigenetic and constructivist models of evolution (Ho and Saunders, 1984). Evolutionary theory formed a crucial part of the matrix of ideas from which child and developmental psychology emerged in the late nineteenth century, and one of the themes which runs throughout this book is the dialectical and epigenetic relationship between representation and adaptation. The reference to evolutionary theory is one sense in which the 'naturalistic' part of the book's subtitle can be read.

Although my main concern is with the ontogenesis of language and representation, part of my argument is that ontogenetic, phylogenetic and historical time-scales are interwoven in concrete human development, and that such a perspective offers, as Piaget and Vygotsky understood, a *tertium quid* between nativism and environmentalism. I also try to show, however, how Piaget's and

Vygotsky's own formulations were compromised by their acceptance of a complex of presuppositions regarding 'evolution', 'culture' and 'progress' which were current at the turn of the century; and to offer a more adequate account.

Another sense in which the term 'naturalism' is here employed is summarized in the following quotation from Bhaskar (1979: 3):

> *Naturalism* may be defined as the thesis that there is (or can be) an essential unity of method between the natural and the social sciences. It must be immediately distinguished from two species of it: *reductionism*, which asserts that there is an actual identity of subject matter as well; and *scientism*, which denies that there are any significant differences in the methods appropriate to studying social and natural objects, whether or not they are actually (as in reductionism) identified.

The notion of a socio-naturalistic approach to developmental psychology thus involves both a theoretical proposition regarding the specific nature of the object of inquiry—one which is *both* natural and social—and a methodological proposition regarding the appropriate means for conducting the inquiry. The data presented in Chapter 5, which were the result of collaborative experimental work, form an important but not uniquely decisive part of this inquiry. As van Geert (1983: 214) argues, 'The "sense" of [an] experiment—or, more precisely, its Saussurean "value"—depends on its place in the structure of events and concepts, which is clearly theory dependent. It may even be questioned whether the referent can remain unaffected when ... basically different views on the nature of psychological reality are concerned.' It is because I believe that these experimental data can only adequately be made sense of in the context of a theory that I have postponed presentation and discussion of them until after the critical discussion of existing theories, in the course of which my own theoretical approach, which I have dubbed 'the materiality of representation', will I hope have become clear.

Although critique is essential to the overall aim of the book, that aim itself is more ambitious: to outline a general, developmental theory of language and cognitive representation. For better or worse, this aim dictates the interdisciplinary nature of both the subject matter treated and the approach I have taken to it. In Chapters 1 and 2, the problem of representation is explored from philosophical, semiotic and linguistic points of view. A pragma-semiotic account of representation is presented, and its consequences for psychological theory are explored. Chapters 3 and 4 examine evolutionary, psychobiological and cognitivist approaches to language and representation, and advance a theory of human representational development designated as 'epigenetic naturalism'. Chapter 5 theorizes the notion of 'context' in terms of discourse-pragmatic, semiotic and cognitive processes, and presents a specific account of language acquisition, focussing upon word meaning.

The adoption of an interdisciplinary perspective poses obvious problems regarding both the presentation, and the intended audience, of a book. I have presented what is possibly 'new information' for many psychologists—particularly recent developments in the philosophy of language, including post-structuralist accounts—in a way that, I hope, is accessible. I have also tried not to presuppose too much specific psychological knowledge, in the hope that this book will find an audience amongst philosophers, linguists and non-psychological social scientists as well. Although this is not a textbook, I hope it contains sufficient review material, presented in a sufficiently clear manner, to serve as a text for critical appraisal in advanced undergraduate courses, as well as contributing to an ongoing dialogue with other researchers in developmental psychology and psycholinguistics.

Acknowledgements

In the course of thinking about and writing this book, I have greatly benefited from discussions with many colleagues and friends, some but by no means all of whom are referenced in the text. I mention here only those who have been of particular importance in making this book possible at all, but my gratitude extends to many others as well.

Gordon Wells first gave me employment as a research assistant on the project 'Language Development in Pre-School Children' at the University of Bristol. I could have had no better apprenticeship to the fascination and illumination to be gained from research with young children; nor to the problems and pitfalls which beset it. Valerie Walkerdine also worked on that project, and as well as collaborating with her then in research which was deeply formative of my later work, I have since enjoyed many years of her friendship and intellectual companionship. The first experiment reported in Chapter 5 is based upon an earlier investigation which she and I jointly conducted. Norman Freeman, with whom I jointly designed and executed, at the University of Bristol, both experiments reported in Chapter 5, taught me much of what I know about translating theories into experimental designs; and was as patient as anyone could be expected to be with my tendency subsequently to translate them back again. George Butterworth has given me consistent encouragement in the writing of this book; and both he and the editorial staff of the Harvester Press have borne patiently its long gestation and multiple revisions. Wolfgang Klein and Piet Vroon provided the ideal working conditions which enabled me to begin and finally to complete the manuscript. Hedwig Lohmann combined forbearance and support in perfect proportion. My thanks to all of them.

About half of the material in this book has appeared in previous publications, but has been substantially revised for the present text. Parts of Chapter 1 and 3 appeared in A. Lock and C. Peters (eds.) *The Handbook of Human Symbolic Evolution* (© 1988 by Oxford University Press). Parts of Chapter 4 appeared in M-W. Ho and P.

Saunders (eds) *Beyond Neo-Darwinism: An introduction to the new evolutionary paradigm* (© 1984 by Academic Press, Inc.). Parts of Chapter 5 appeared in T. Seiler and W. Wannenmacher (eds) *Concept Development and the Development of Word Meaning* (© 1983 by Springer-Verlag); and in G. Hoppenbrouwers, P. Seuren and A. Weijters (eds) *Meaning and the Lexicon* (Foris Publications, 1985). Reproduction of copyright material is by permission of the publishers.

The quotation appearing on pp. 56–57 is from "Introductory essay: notes on conversation" by John Searle, 1986, in D. Ellis and W. Donohue (eds) *Contemporary Issues in Language and Discourse Processes*, pp. 16–17 © 1986 by Lawrence Earlbaum Associates, Inc., reprinted by permission.

The preparation of this book was financially supported by the Economic and Social Research Council (GB); the Max-Planck Institute for Psycholinguistics, Nijmegen; and the research institute of the Faculty of Social Sciences of the State University of Utrecht.

Nijmegen, November 1987

1 Theories of the Sign

Too many things are signs, and too different from one another.

Umberto Eco (1984: 18)

"How beautiful the world is, and how ugly labyrinths are," I said, relieved.
"How beautiful the world would be if there were a procedure for moving through labyrinths," my master replied.

Umberto Eco (1983: 178)

No form of semiotics can exist other than as a critique of semiotics.

Julia Kristeva (1986: 78)

Introduction

All representation presupposes some medium within, upon or by means of which the message, intended or unintended, is carried. Metaphors such as 'imprinting' and 'etching' abound in the literature relating to cognition, perception and memory, testifying both to the naïve realism underlying the still-predominant empiricist approaches to the problem of representation; and to the ancient intuition that the sign is a 'mark' or 'trace' which permits both universalization and individuation. The material substrate of representation, however, is not restricted to the medium as such, whether this is conceived of as paper and ink, computer hardware, or the neural architecture supporting mental and linguistic processes. It is also to be sought, at least in the case of human symbolic processes, in the objective, systemic properties of languages and sign systems, as manifested in spoken utterances, written texts, rituals and so forth; in the social practices and relations which are realized and sustained by linguistic communication; and in the tools, artefacts and enduring symbols which represent the technical-practical capacities of the society, the results of past labour and the organizing values of the culture.

Evolutionary and developmental theories seek their explanatory principles in the dynamic relationships between the life cycle of the organism and the (changing) environment. In the case of human development, the relevant material environment includes as a decisive factor the representational environment which is the product of symbolic, constructional and social practices. An adequate theory of human development must, therefore, address the nature of representation, signification and symbolic behaviour.

The notions of 'representation', 'symbolization' and 'signification' (and others such as 'denotation', 'designation', 'reference' etc.) can only satisfactorily be defined in a recursive (self-referential) fashion; they 'stand for' the relation which they name, that of 'standing for' (hence the medieval definition of the sign as 'aliquid stat pro aliquo'). The notion, like the relation, is deceptive in its seeming simplicity. Even the most apparently transparent representation, such as a photograph, depends for its interpretation upon a complex and culturally specified repertoire of norms and beliefs, themselves inscribed within a regulative apparatus for the reproduction of a social order (Barthes, 1973; Berger, 1980; Sontag, 1982). There is no such thing as 'absolute' likeness, and no ultimate standard of verisimilitude against which representations may be judged, in the manner in which nineteenth-century scientists judged length by reference to absolutes embodied in bars of metal at Greenwich or Paris. Yet, as we shall see, this does not mean that there are *no* canons or rules governing the structure of representations, nor that such canons are entirely independent of the nature of the world and of our own species-being within it.

An even greater degree of opacity attaches to symbolic entities, such as linguistic signs, whose relation to what they stand for, in the world or in the mind of the languager user, appears to be grounded not at all in resemblance or likeness, but solely in formal rules and arbitrary conventions. As we shall also see, however, the old argument between 'nominalism' and 'realism' has refused, despite the best efforts of linguists and philosophers, to accede to the repeated death-notices served upon it.

The overall theoretical approach adopted in this book can be characterized by the slogan 'the materiality of representation'. I shall argue that representation, including 'mental' representation, cannot be understood as a secondary structure superimposed upon an innocent world of form and matter—'things' are not only not as they seem, as we already know from modern physics, but the world of things is also representational in its material structure. Accustomed as we are to thinking of 'things' as material, and rules, relations and meanings as derivative and (merely) mental, it requires a considerable initial effort to grasp the proposition that representation is constitutive, not just of our understanding of the world, but of the world itself as it presents itself to our understanding.[1] This is, however, the essence of the argument which I shall develop, beginning with an examination of the sign and signification as understood in classical and modern semiotic theory.

Thought and the sign in classical philosophy

Semiotics, or the science of signs and sign-systems, had its origins in the classical and scholastic study of the arts of Logic, Rhetoric and Poetics. The earliest usage of the term 'sign' (Greek = *semeion*) in a philosophical context apparently derived from Hippocratic or Asklepiadic medicine, with its concern for symptomatology and inferential diagnostics. The classical Greek notion of the sign referred to 'something immediately evident which leads to some conclusion about the existence of something not immediately evident' (Eco, 1984: 31), as in the well-known example of 'smoke signifies fire'. A close relationship between the concept of signification, and those of evidence and inference, can be found in the writings of Aristotle. His successors the Stoics elaborated this into a theory of strict implication: for Sextus Empiricus (cited by Eco, 1984: 15) the sign was 'a proposition constituted by a valid and revealing connection to its consequent'. Thus, for the Greeks the study of signs was part of the general inquiry into the operations of reason and the possibility of knowledge; through and in signs, nature revealed itself to the intellect. The Greek theory of the sign was not concerned with intentional or communicative signification, nor in the main with language, except inasmuch as the latter serves as the vehicle for propositions.

An important exception to this general observation may be found in the Aristotelian notion that 'thoughts' may be considered as representatives of 'things', and that language in its turn is the sign or mark of thought. Aristotle said that words are signs of 'affections of the soul', which he in turn considered to be mental images or copies of objects. Eco (1984) maintains that the Greeks, because of their emphasis upon 'natural' signs and their place in logic, did not regard language as a species of signification, and that this proposal originated in the fifth century with St Augustine; but in *De Interpretatione* Aristotle writes that 'Spoken words are the signs of affections of the soul, and written words are the signs of spoken words. Just as all men have not the same writing, so all men have not the same speech sounds, but the affections of the soul which these signify are the same for all, as also are those things of which our experiences are images.'[2]

This view persisted until the close of what Foucault (1970) terms the 'Classical' age in the late eighteenth century. Thus, Locke maintained that: 'That then which words are the marks of are the ideas of the speaker: nor can anyone apply them as marks, immediately, to anything else than the ideas that he himself hath.'[3] It was this notion, too, that underlay the well-known argument of Descartes, taken up by his successors the Port-Royal grammarians, regarding the uniqueness of the human faculty of reason. Descartes wrote:

> If there were machines which bore a resemblance to our body and imitated our actions as far as it was morally possible to do so, we should always have two very certain tests by which to recognize that, for all that, they were not real men. The first is, that they could never use speech or other signs as we do when placing our thoughts on record for the benefit of others. For we can easily understand a machine's being constituted so that it can utter words, and even emit some responses to action on it of a corporeal kind, which brings about a change in its organs ... But it never happens that it arranges its speech in various ways, in order to reply appropriately to everything that may be said in its presence, as even the lowliest type of man can do. And the second difference is, that although machines can perform certain things as well as or perhaps better than any of us can do, they infallibly fall short in others, by the which means we may discover that they did not act from knowledge, but only from the disposition of their organs. For while reason is a universal instrument which can serve for all contingencies, these organs have need of some special adaptation for every particular action. From this it follows that it is morally impossible that there should be sufficient diversity in any machine to allow it to act in all the events of life in the same way as our reason causes us to act.[4]

By one of history's odd ironies, this Cartesian line of reasoning is associated in contemporary cognitive theory with a computational theory of mind, cast in terms of specially adapted 'mental organs', of precisely the sort that Descartes thought to demonstrate as unrealizable. In any event, the issues that he addressed—whether reason is 'integral' or 'modular'; whether there is some specific quality to reason (call it soul, intentionality, or whatever) which sets it apart from mere calculation and reaction; whether the language faculty is species-specific, and whether its exercise is an all-or-nothing affair, or a matter of degree—are with us today as much as then. Such questions, in fact, constitute not only a persistent thread in the philosophy of language, but also the fundamental matrix for the repeated attempts to found an empirical science of psycholinguistics. Almost two hundred years after Descartes formulated his theory of language and mind, in post-Revolutionary France, the physician Itard concluded (following Locke and Condillac but against Descartes), from his experimental education of Victor, 'the wild boy of Aveyron', that 'the moral superiority said to be *natural* to man is only the result of civilization, which raises him above other animals by a great and powerful force' (cited in Lane, 1977: 129.)

For Descartes, and for the entire tradition of 'Classical' philosophy including both rationalism and empiricism, the possibility of knowledge was given by, on the one hand, intuition, and on the other hand inference or judgement. Representation—the subject's 'ideas' of things—guaranteed the first and fundamental basis of knowledge, that of intuition. Intuition was initially deployed by Descartes in relation to the subject—the Cogito. It should not be forgotten that the seventeenth century was an age of doubt, in which the fragile sense of individuality brought into being by the Renaissance was riven by a sense of turmoil and contradiction mirroring the social, political and religious conflicts engendered by the bourgeois and scientific revolutions. This crisis of individuality was expressed by the English poet and Parliamentarian Andrew Marvell, for whom the Individual:

> nak'd and fierce doth stand
> Cuffing the thunder with one hand;
> While with the other he does lock,
> and grapple with the stubborn Rock.
> (The Unfortunate Lover)

Descartes sought and found his Rock in the certainty that, as reason proceeds outwards from the 'I', that constitutive 'I' must both assuredly exist, and be the starting point of all knowledge. In so doing, he both initiated that 'methodological individualism' which stands at the heart of the human sciences and their discontents; and secured the foundations of the tradition of Western philosophy whose antinomies (subject/object, individual/society, nature/culture, mind/reality) are the most evident *loci* of these very discontents. More immediately, however, Descartes opened the space (between intuition and judgement) within which empiricism and rationalism contested the claims of experience and inheritance.

As both Foucault (1970) and Putnam (1981) have, in different ways, remarked, the Classical (pre-Kantian) epistemologies proceeded by an identification of the sign with representation, the latter conceived in terms of resemblance or similitude. It was the capacity of thought (representation) to represent itself in the signs of language that furnished Man with the means to reason, and indeed to 'prescribe rules for his judgement, which is logic, for his discourse, which is grammar, for his desires, which is ethics'; all of which, in true Cartesian fashion, derive from 'a common source ... this sole centre of all truths is the knowledge of his intellectual faculties' (Destutt de Tracy, cited in Foucault, 1970: 85). Language, in these theories, *unites* representation with itself, for, in grammar, discourse and judgement are united in the proposition, the universal form and representative figure of thought.

This framework was accepted by both rationalists, who believed the representations underlying intuitions and judgements to be innate; and by empiricists, who believed that ideas derived (via 'abstraction') from the imprint of experience, more complex judgements being developed through the association of ideas and sensations (so that, for example, causality might be reduced, as in Hume, to contiguity). These differences were, however, secondary to the common acceptance of the notion of representation as resemblance, whether evoked in intuition through the affections of the senses, or invoked in judgement through the application of the laws of logic. In this respect, it is misleading to consider empiricism and rationalism as doctrines which were simply opposed; and more illuminating to regard both as variants of a single 'empirio-rationalist' paradigm.

Post-classicism: the Kantian turn

The eventual crisis which led Kant to formulate his critique of empirio-rationalism had its roots in precisely this notion of representation as resemblance: or, as Putnam terms it, the 'similitude theory of reference'. Putnam's account is so clear that it is appropriate to quote him here:

> In the seventeenth-century the similitude theory began to be restricted ... Thus Locke and Descartes held that in the case of a "secondary" quality, such as a colour or a texture, it would be absurd to suppose that the property of the mental image is *literally* the same property as the property of the physical thing ... [However] for those [primary] properties (shape, motion, position) which his Corpuscularian philosophy led him to regard as basic and irreducible, Locke was willing to keep the similitude theory ... Berkeley discovered a very unwelcome consequence of the similitude theory of reference: it implies that nothing exists except mental entities ... Berkeley's argument is very simple. The usual philosophical argument against the similitude theory in the case of secondary qualities is correct (the argument from the relativity of perception), but it goes just as well in the case of primary qualities ... To ask whether a *table* is the same length as *my* image of it or the same length as *your* image of it is to ask an absurd question ... *Nothing can be similar to a sensation or image except another sensation or image.* Given this, and given the (still unquestioned) assumption that the mechanism of reference is similitude between our "ideas" and what they represent, it at once follows that no "idea" can represent or refer to anything but another image or sensation ... And if you can't think of something, you can't think it exists (Putnam, 1981: 57–59).

It was against the background of this impasse that Kant inaugurated what he termed his 'Copernican Revolution' in philosophy. Kant accepted, with Descartes, Locke and others, that the possibility of knowledge was to be located in the space between intuition and judgement; he also recognized that Berkeley's extension of Locke's arguments rendered the attempt to derive judgement from intuition, by way of experience, unworkable. Kant, in fact, took Berkeley's argument further, in arguing that concepts cannot be images at all, for the simple reason that images can never comprehend the diversity-in-unity of the totality of objects (whether real *or* mental) subsumed under the concept: 'No image would ever be adequate to the concept of a triangle in general. It would never attain that universality of the concept which renders it valid of all triangles, whether right-angled, obtuse-angled, or acute-angled; it would always be limited to a part only of this sphere'

(Kant, 1929 [1787]: 182).[5] An image, then, is inadequate as a *universal* (i.e. conceptual) representation not only of a class of objects, but also of a class of images. The argument from the relativity of perception leads inexorably to the conclusion that concepts are not images, and that intuitions are not based upon images: the similitude theory of reference must be rejected.

While Berkeley used the argument from the relativity of perception to propose a theory in which the object is itself subjective—hence the name of his philosophy, 'subjective idealism'—Kant's use of the argument took an entirely different turn.[6] Accepting that no perspectively located representation can be adequate to its object, and that *all* representation is perspectively located—ultimately, to use the much later terminology of Wittgenstein, in a 'human form of life'—Kant argued that a representation of the object 'in itself'—that is, viewed from *no* perspective or *every* perspective—is unattainable, since no perspectively located 'copy', however ideal, can be adequate to the plenitude of the object.

There is a considerable degree of controversy regarding the Kantian notion of the unknowability of the object in itself, which is frequently held to be a metaphysical proposition. However, for the purposes of this exposition, it can be supposed that Kant meant that objects in themselves are unknowable (unrepresentable) in an approximative, vanishing or asymptotic sense; it isn't so much that there is some essentially unknowable quality of objects, as that we can only approach and apprehend them perspectively. Similarly, Marková (1982) understands the Kantian theory of mental representation (see below) to imply that, in thought or discourse, we think or talk, not about 'the world' but about our representations of it; while on the interpretation which I am advancing, what Kant meant was that we can think or talk about the world only as it presents itself to our understanding (representations).

This conclusion led Kant to a reformulation of the relation between knowledge and experience, in which the (as he considered it) misleading focus of empirio-rationalism upon the 'surface' appearance of objects, with the concomitant concept of representation as resemblance, was replaced by a *critical* concept of representation as a relation of *adequation* between judgement and its object.[7] Kant believed profoundly in science and progress, and in the experimental–observational method that had yielded a rich harvest of knowledge in the short time since its principles were first enumerated by Francis Bacon. Yet, as he argued, 'Natural science ... in so far as it is founded upon *empirical* principles ... [has] learned that reason has insight only into that which it produces after a plan of its own, and that it must not allow itself to be kept, as it were, in nature's leading strings, but must itself show the way with principles of judgement based upon fixed laws, constraining nature to give answer to questions of reasoning's own determining' (Kant, 1929: 19–20).

It was clear to Kant, then, that new knowledge—better understanding—could arise only from an interaction between judgement and its object: and that this object ('representation'), though it must be *open* to experience, and indeed in some sense itself *constitutes* experience, could not be derived *from* experience through abstraction. As he famously wrote:

> There can be no doubt that all our knowledge begins with experience. For how should our faculty of knowledge be awakened into action did not objects affecting our senses partly of themselves produce representations, partly arouse the activity of our understanding to compare these representations, and, by combining them or separating them, work up the raw material of the sensible impressions into that knowledge of objects which is entitled experience? In the order of time, therefore, we have no knowledge antecedent to experience, and with experience all our knowledge begins.
>
> But though all our knowledge begins with experience, it does not follow that it all arises out of experience ... even our empirical knowledge is made up of what we receive in impressions and of what our own faculty of knowledge (sensible impressions serving merely as the occasion) supplies from itself ... such knowledge is entitled *a-priori*, and distinguished from the *empirical* which has its sources *a-posteriori*, that is, in experience (Kant, 1929: 41–42).

For Kant, the 'a-priori' was not a subset of ideas in general (Kant broke with the empirio-rationalist problematic within which it might be argued that some ideas were innate and others derived from experience), but the condition of possibility of *all* representation (whether of objects, in intuition, or of ideas and hence of thought). Thus, instead of deriving judgement from either intuition or representation, singly or in any combination, Kant reversed the problem and derived the conditions of possibility of representation from the nature and operations of judgement.

The understanding has thus far been explained merely negatively, as a non-sensible faculty of knowledge. Now since without [sensation] we cannot have any intuition, understanding cannot be a faculty of intuition. But besides intuition there is no other mode of knowledge except by means of concepts. The knowledge yielded by understanding, or at least by human understanding, must therefore be by means of concepts, and so is not intuitive, but discursive. Whereas all intuitions, as sensible, rest on affections, concepts rest on functions. By "function" I mean the unity of the act of bringing various representations under one common representation ... the only use which understanding can make of these concepts is to judge by means of them. Since no representation, save when it is an intuition, is in immediate relation to an object, no concept is ever related to an object immediately, but to some other representation of it, be that other representation an intuition, or itself a concept. Judgement is therefore the mediate knowledge of an object, that is, the representation of a representation of it. In every judgement there is a concept which holds of many representations, and among them of a given representation that is immediately related to an object ... Now we can reduce all acts of understanding to judgements, and the *understanding* may therefore be represented as a *faculty of judgement* ... The functions of the understanding can, therefore, be discovered if we can give an exhaustive statement of the functions of unity in judgements (Kant, 1929: 105–106).

We shall not dwell here upon the particular forms (the Table of Judgements, the synthetic a-priori and so forth) in which Kant cast his proposed transcendental logic. What is important for current purposes is the way in which he transformed the problem of knowledge. Before Kant, attaining knowledge was conceived as a matter of the faithful reduplication of the surface of nature *through* representation; after him, it became a matter of harnessing empirical inquiry to judgement in order to reconstruct, *in* representation, the hidden structures underlying phenomena. Kant thus both inaugurated the *constructivist* epistemological stance, which was later to be elaborated by Piaget, and recast in modern terms the notion that scientific knowledge consists in going *beyond* appearance to disclose the 'deep structure' of the world. In this respect, at least, Chomsky's generative linguistics is more Kantian than Cartesian. And in this respect, too, Marx was simply following Kant when he wrote that: 'all science would be superfluous if the outward appearance and the essence of things coincided'.[8]

A further, equally profound, consequence of Kant's inversion of the priorities and claims of judgement and representation was that the 'functions of unity', which guarantee the coherence of knowledge, are displaced from representation to judgement.

Representation, the locus in the Classical theories of the unification of reason with its object, acquires a contingent, rather than necessary character, since only in the case of intuitions (which, as Kant emphasises, are not the basis for concepts and judgements) is representation 'immediate'. Discursive concepts cease to be bound to the world by representation, and hence discourse is severed (or liberated) from representation. This is how Foucault puts it: 'whereas before it was a question of establishing relations of identity or difference against the continuous background of similitudes, Kant brings into prominence the inverse problem of the synthesis of the diverse ... Kant avoids representation itself and what is given within it, in order to address himself to that on the basis of which all representation, whatever its form, may be posited' (Foucault, 1970: pp. 162, 242).

In Kant's theory, representation itself, though it ceases to be central to epistemological considerations, is extensively considered as a *psychological* problem; Kant in fact addresses, for the first time, the problem of *mental representation* (Vorstellung) as it is understood in contemporary cognitive science. It was, of course, necessary for Kant, in rejecting the similitude theory of reference, to provide some alternative account of how concepts are articulated with experience. This he did with the celebrated theory of the 'schemata' of understanding. This is how Kant, after dismissing the possibility that an image, or copy, can be adequate to the universal character of a concept, distinguishes the 'schema' from the image: 'Indeed it is schemata, not images of objects, which underlie our pure sensible concepts [such as that of the triangle] ... Still less is an object of experience or its image ever adequate to the empirical concept, [which] stands in immediate relation to the schema of imagination, as a rule for the determination of our intuition, in accordance with some specific universal concept. The concept "dog" signifies a rule according to which my imagination can delineate the figure of a four-footed animal in a general manner, without limitation to any single determinate figure such as experience, or any possible image that I can represent *in concreto*, actually presents' (Kant, 1923: 182–183).

As to the actual mechanisms and origins of schemata, Kant held these to be a mystery. The concept of the schema, however, has exerted a powerful influence upon cognitive psychology from its inception, in the work of Herbart, through Wundt, Bartlett and

Piaget to contemporary theories of memory, comprehension and language acquisition (e.g. Bartlett, 1932; Abelson, 1981; Nelson, 1983; Schank, 1982; for a historical overview, see Kessel and Bevan, 1985).

Although Kant was not a philosopher of language, his investigations profoundly influenced the development of theories of language. Firstly, his separation of representation from language and discourse implied that language itself could no longer be seen as a transparent tool for the revelation of both nature and reason, but must rather be studied as an object of investigation in its own right. Language, divested of its capacity to unify representation with its object, was henceforth subject to the higher claims of empirical investigation: language was naturalized, in the same movement that de-naturalized representation.

Secondly, Kant and, to a much greater extent, his successors such as Hegel, Goethe, Herder and von Humboldt, approached natural objects from an evolutionary-developmental point of view (see Marková, 1982). For this reason, nineteenth-century philology and linguistics approached language in terms of origin, evolution and comparison; the revolutionary implications of Darwinism coinciding here with the pre-occupations of the German romantic culture-philosophy. The emphasis in this tradition was, on the one hand, upon language as expression, both of individual subjectivity and of cultural identity (the 'genius of the race'); and, on the other hand, upon the derivation of modern from archaic languages. The romantic-expressivist tradition based its criticism of Kant upon the latter's 'psychologism'—his grounding of the 'synthesis of apperception' in a transcendental consciousness divorced from both society and sign use. While the Kantian line of reasoning was subsequently pursued in Husserl's phenomenology, the romantic-expressivist critique of it found its later direction in hermeneutic and historicist philosophies.

The historicist, diachronic aspect of the post-Kantian tradition was clearly manifest in the work of the 'neo-grammarians', which was the target of Saussure's structuralist critique. It was to surface again in Bakhtin's double critique of structuralism and expressivism, and in Prague School linguistics. Hermeneutical concerns also later re-emerged in the work of, for example, Mead, Whorf, and the later Wittgenstein. More immediately, however, nineteenth-century linguistic theories were eclipsed by the rise of structuralism and positivism, in which socio-historical speculation gave way to systematic, formal analysis and a renewed attention to the logical properties of language. The 'liberation' of discourse from representation thus yielded to its enslavement to the sign, which inherited in twentieth-century linguistic theory the contingent character accorded by Kant to representation, recast as the concept of 'arbitrariness'.

Modern signs: Peirce, Frege, Saussure

In the late nineteenth and early twentieth centuries, the study of language diverged into different disciplinary pathways, resulting in a division of labour which, despite the rise of new 'interdisciplines' such as cognitive science, persists to this day. Philosophers, in search of new and more powerful logics, have explored and evaluated natural language expressions against the background of their preoccupation with formal, propositional truth. Linguists, at least until recently, eschewed such problems in favour of an essentially descriptive enterprise, seeking to systematize the facts of language structure independently of its referential or communicative functions. Psychologists have looked to linguistic processes for exemplification of the general principles underlying learning, perception and cognition. Sociologists and anthropologists have viewed langauge as a cultural form reflecting or implementing the ideological and 'superstructural' reproduction of social relations (see Williams, 1977, for a critical discussion).

In spite of these distinctions, however, 'modern' linguistic and semiotic science, as essentially adhered to by all these currents of thought, consists largely in the elaboration and critique of the crucial insights of three figures, two philosophers and a linguist, whose ideas—more or less contemporaneously developed around the turn of the century—have profoundly influenced our understanding of the sign and signification.

Peirce

In many ways, the work of Charles Sanders Peirce (1839–1914) may be seen as simultaneously a return to the problematic which inspired Kant, and an attempt to save the intricate systematics of the Scholastic theory of signs. Peirce's distinction consists principally in that he constructed a theory of signs which treated linguistic and non-linguistic semiotic processes within a unified framework. Problems of the nature and ground of knowledge—epistemological questions—were also addressed by Peirce, in far greater depth than had previously been attempted within the context of the theory of signs. Peirce and Frege (discussed below) can thus fairly be said to have inaugurated the shift from a transcendental to an analytic philosophy of mind which set the scene for many of the most important developments in twentieth-century philosophy.[9]

Peirce's theory consists in essence of the juxtaposition of two sets of threefold, or triadic, distinctions; one concerning the *mechanism* of the sign, and its internal relations; the other concerning the *nature* of different types of sign according to the manner in which the internal relations operate. All signs share a common structure, in that they depend upon and are defined in terms of the relations between three elements. The first element, for which Peirce himself frequently used the simple term 'sign', consists in the perceptible unit (for example, a group of phonemes, a printed word, a traffic light) which conveys the meaning of the sign. McNeill (1979), whose account I shall largely follow, therefore calls this element the sign vehicle (see also Bates, 1979). Hereafter, I shall refer to this element as the *signifier*. The second element of the sign is the *object* which the signifer can be interpreted as 'standing for': which need not be either a physical or a real object, and may indeed be another sign. The third element of the sign is the *interpretant* of the sign, by virtue of which the signifier signifies the object for a subject. The interpretant is at once the *cause* of the signifying relation between the signifier and the object, in that the relation between the interpretant and the object duplicates the relation between the signifier and the object; and the *result* of the act of signification, in that the interpretant is the 'significate outcome' (in the mind of the subject) of the interpretation of the signifier as signifying the object.

According to Peirce, the interpretant is therefore *also* a sign, albeit of a mental or cognitive nature. The following quotations from Peirce (taken from McNeill, 1979:40) make this clear: 'A sign is only a sign *in actu* by virtue of its receiving an interpretation, that is by virtue of its determining another sign of the same object ... A *sign* is a representamen of which some interpretant is a cognition of mind.' For Peirce, then, the sign-relationship can be understood only if the interpretive nature of signification, involving a perceiving and/or cognizing subject as well as a 'signifier' and a 'signified', is taken into account.

It should be noted, though, that the recognition accorded to the subject in the *interpretation* of the sign does not require that the *production* of the sign be oriented to the needs of the interpreter, nor indeed that the sign be intentionally produced. As McNeill (1979: 42) points out, 'Peirce casts the discussion of signs in the context of a general epistemological theory in which knowledge is gained through sign interpretations ... This function of the sign in the acquisition of knowledge introduces an asymmetry wherein the recognition of signs is favoured over the construction or emission of signs.' Thus, Peirce's semiotic epistemology, although it can be said to be 'communicative' inasmuch as the interpretation of signs frequently presupposes the existence of a sign-using community, embraces a larger set of cognitive processes than those involved only in communication. In this respect, it differs from the 'code-oriented' semiotics of Jakobson and others. Moreover, because Peirce's theory is one of signs in general, as opposed to the narrower category of intentionally produced signs, it also embraces a larger category of signs than those which can be called 'representational'; an issue which I analyse more fully in Chapter 2.

Relations between signifier and object (and hence also between interpretant and object) may be of three kinds. In the case of an *indexical* sign, the signifer is related to the object by virtue of a real relationship (such as physical causation, or part–whole relations) obtaining between signifier and object, or a relationship such that the signifier directs attention by 'blind compulsion' to the object. Thus, the symptoms of a disease, the smoke of a fire, the tip of an iceberg and a pointing finger are all examples of indexical signifiers. In the case of an *iconic* sign, the signifier is related to the object by virtue of its resemblance to the object, as is the case with pictures, images or diagrams. In the case of *symbolic* signs, the signifier is related to the object by virtue of a rule or convention, as is the case with a word such as 'dog', which signifies by virtue of a complex of linguistic rules and conventions, rather than by virtue either of necessity (indexicality) or similarity (iconicity).

Much has been written about the problems and issues associated with Peirce's typology of signs, and I shall restrict myself here to three comments. First, it is frequently the case that different signifying relations may be combined in a single sign. For example, as Bates (1979) points out, the footprint of an animal is iconically related to the shape of the animal's foot, and is simultaneously an indexical signifier of the presence of the animal at this place at a previous time. Equally, although language is an essentially symbolic signifying system, it also presents aspects of iconicity (as in the temporal ordering of narrative sequences) and indexicality (as in the contextual determination of the reference of deictic terms, such as 'this', 'I' and 'here').

Second, the nature of symbolic signs as being governed by convention is often taken to imply that all conventions in symbolic systems such as language are arbitrary, an assumption which is largely grounded in the work of Saussure, which is discussed below. Not only have the iconic aspects of language, as McNeill (1979) points out, frequently been underemphasized, but the *motivation* of certain conventions by functional considerations has also frequently been underplayed (Paprotté and Sinha, 1987).

The third issue is more complex, and relates to the extent to which Peirce's theory applies only to the internal workings of signs, or can be seen also to solve the problem of how signs are used to refer in and to the real world. I have presented here a largely 'internalist' account, emphasizing that the process by which signs are interpreted is one involving further signifying relations, between interpretant and object (bear in mind that not only is the interpretant also a signifier, but the object may itself also be a sign). Peirce himself did frequently use the terms 'refer' and 'reference', but he lacked a theory of reference, and in this respect his most important contribution may be seen precisely in producing a theoretical apparatus capable of comprehending the recursive and pluri-significant nature of semiotic processes.

Frege

In order to move beyond signification to reference, we must next consider the work of Gottlob Frege (1848–1925). Frege was a logician and mathematician, as well as a philosopher of language, and his approach to linguistic signification was motivated by an attempt to understand how the substitution of one referring expression for another preserves or does not preserve the truth of a proposition. Traditional logics distinguished between propositions expressing 'analytic' truths—those which, in modern terms, are necessarily true in all possible worlds, such as 'all humans are mortal'; and those expressing 'empirical' truths, true contingently within our knowledge and experience, such as 'Mount Everest is the highest mountain on earth'. The notion of analyticity remains problematic (see Kripke, 1980; Putnam, 1981), but Frege (1892) started his analysis by observing that the same individual entity may appropriately be referred to on different occasions using different expressions. To use his own example, the expressions 'the morning star' and 'the evening star', while unrelated by analytical logical entailments, share a common referent (Venus).

This example is in some respects a special case, but the phenomenon of the multiplicity of possible referring expressions for a given entity is a general one (Brown, 1958). Thus, Frege distinguished between the *sense* (Sinn) of a referring expression and its *reference* (Bedeutung): two expressions, then, may have quite different senses (e.g. 'my neighbour' and 'the doctor'), allowing for quite unrelated inferences (e.g. 'lives next door', 'is medically qualified'), while sharing the same referent. This phenomenon also underlies the *informative potential* of language, by permitting multiple predications of the same referent: e.g. 'my neighbour is a doctor'.

On the other hand, while the same expression (e.g. 'the doctor') may be used to refer to different *individuals*, it is also the case that expressions which are analytically identical in their sense (e.g. 'doctor' and 'medically qualified person'—NB the expression 'doctor' is used here *in the sense of* 'medical doctor') have identical *scope* of reference; they refer to the same *class* of individuals. Thus, a general asymmetry exists in the relations between sense and reference: while referring expressions of different sense may refer to the same individual, a referring expression may not be used to refer to an individual if that individual cannot be referred to by an expression with identical sense. Therefore, according to Frege, it is the sense of a term which determines its reference; and, further, the sense of a complex expression (and hence its reference) is determined by the senses of the constituent expressions into which it may be analysed.

Our understanding of language, then, depends upon our grasp of sense, or meaning, since it is meaning which determines and delimits appropriate reference, and not reference which determines meaning. If this be the case, what do senses consist of? Frege's answer was that 'sense' is an intrinsic and objective property of language, which is grasped or apprehended by the cognitive activity of the subject. Senses, then, and their role in cognition, may be regarded in much the same way as Peirce's interpretants: as signs, but, in this case, signs which are proper to language and which underlie the mechanisms of linguistic reference. However, these signs are, on the Fregean view, *real*, and (unlike Peirce's interpretants) not psychological constructs; they are properties of the world (or at least of the world of language), and not of the (individual) mind, whose task is to 'grasp' and entertain them in cognition.

Saussure

Both Frege and Peirce introduced signification into their respective re-workings of Kantian epistemology, as a means to secure judgement to its object by further specifying the nature and mechanisms of what Kant called 'mediate knowledge'. Both theories also constituted a kind of return to Kantian universalism, a re-affirmation of the necessity of truth as the foundation of philosophy, in reaction against the relativizing tendencies of the romantic-expressive theories of linguistic, cultural and mental difference. In contrast, the theory of the sign proposed by Ferdinand de Saussure (1857–1913)—who is frequently cited as the founder of modern linguistic science—departed both from the Kantian problematic of justified knowledge, and from the romantic-expressive evolutionary and comparative tradition. Saussure's work, published posthumously from the collated lecture notes of his students as the *Cours de Linguistique Générale* (Saussure, 1966 [1915]), has had a tremendous influence, not merely within linguistics, but upon the human sciences as a whole, since in this text he lays out and exemplifies the general principles of the method and philosophy which has come to be known as 'structuralism'.

Before Saussure, linguistics had been dominated by historical and philological concerns. Saussure, on the contrary, insisted that the foundations of linguistic science consisted in the study of the *synchronic* structure of a particular language at a given time, and that the *diachronic* study of language change had necessarily to be secondary to this. Thus, rather than viewing linguistic signs in terms of their origins and historical transmutations, he sought to define the rules governing their lawful combinations and interrelations within a language. Saussure distinguished between *la parole* (speech), consisting of the actual speech acts of language users, and *la langue* (language), the linguistic system adhered to by a community of language users and regulating their speech conduct (parole). The object of linguistic science is *langue*, the facts of language structure, and its goal is their systematic description for a given language.

According to Saussure, a linguistic sign consists of a coupling of a *signifier* with a *signified*. The signifier is the acoustic or graphic 'mark', and the signified is the meaning or concept which it conveys—the Saussurean 'signified' may thus be thought of as equivalent to Frege's 'sense' or Peirce's 'interpretant'. Saussure, however, was concerned not so much with the inner workings of the sign, nor with its referential function, as with its functioning within the sign-system as a totality. At the level of the individual sign, Saussure argued, there is no necessary connection between the signifier and the signified. For example, the English word 'tree' and the French word 'arbre' signify in the different languages according to different conventions. Further, the realm of the signified—or sense—may be 'cut' differently by the signifiers of different languages. For example, the French word 'mouton' signifies a sense domain which may, according to context, be signified by either of the English words 'mutton' or 'sheep'. For this reason Saussure insisted that the connection between signifier and signified is *arbitrary*.

A linguistic sign, he suggested, may therefore adequately be defined only in terms of the *relations* which it contracts with other signs: linguistic units possess a purely relational identity. Such relations may be of two sorts. *Paradigmatic* relations are contracted between elements which may be substituted for each other in an expression, and *syntagmatic* relations are contracted between elements which may be combined with each other in an expression. These relations are in turned governed by the sets of *contrasts* or *oppositions* which constitute the rule system of a language (langue). Such contrast sets exist at different levels. Thus, the sounds *p* and *b* contrast with each other inasmuch as they may be substituted for one another in some (but not all) phonological contexts, and the permissible and impermissible contexts of substitution are equally governed by the totality of contrasts in the sound system of a language. By the same token, the word *dog* contrasts with the word *cat*, and also with the word *bitch*, and the totality of such meaning contrasts constitutes the semantic system of the language.

Thus, by applying Saussure's structural and contrastive method, we may conclude that the *sense*[10] of a term consists in the meaning relations it contracts with other terms; and, in essence, this remains the basis of much contemporary semantic theory, although contrast in the strict sense of antonymy or 'oppositeness' is not the only relevant meaning relation (Lyons, 1977; Cruse, 1986).

Like Peirce, Saussure conceived of linguistics as one branch of a general science of signs, semiotics or semiology, and his structuralist method to be appropriate for the analysis of all semiotic systems. The structuralist method has been applied by anthropologists to the study of kinship systems, culinary practices and so forth, and the general proposition that social practices may best be seen as signifying and symbolic practices has become central to many areas of social theory. 'Structuralism', as a general methodological premise, has also been extremely influential in cognitive developmental psychology, inasmuch as Piaget's genetic epistemology has frequently been alluded to, by Piaget himself and others, as a 'constructivist structuralism' (Piaget, 1971).

Formalism, structuralism and the semiotic enterprise

So much has been written in recent years[11] regarding structuralism and semiotics that further addition to this weighty literature seems unnecessary. However, a short interlude on some main features of structuralist linguistics may serve to throw into relief a number of issues which recur throughout this book. The most important of these concerns the relations between signification, representation and interpretation. I have argued above that a consequence of Kant's philosophical revolution was the divorce of representation from language and discourse. It is evident from the previous section that, of the three theories discussed, that of Saussure is furthest removed from the traditional guiding concerns of Classical sign theory. In the work of Peirce, the problem of representation is at least implicitly addressed in the orientation of the theory towards interpretation; and arguably also in the inclusion in Peirce's typology of signs the category of the iconic sign. Frege's theory of sense and reference also implies, even if it does not explain, the existence of some *positive* representational content which adheres to the sense of a term.

In Saussurean linguistics, however, such content has been entirely excised through the specification of sense, or 'value', in purely systemic and contrastive terms. This subordination of 'content' to 'form'—together with the subordination of history to structure—was both the most powerful methodological and heuristic innovation of structuralism, and a decisive moment in the struggle to distance linguistics and the human sciences from metaphysics and philosophical speculation. In particular, Saussurean structuralism—like Fregean logical analysis—offered analytical methods untainted (as its advocates saw it) by the psychologism and historicism of the German philosophical and philological tradition; unlike Fregean analysis, it also offered a paradigm for a scientific approach to all aspects of culture and society. The structuralist turn, then, was very much in line with general tendencies in the early twentieth century towards positivist views of science.

The structuralist repudiation of representation as a category amenable to scientific analysis coincided too with the rise of avant-garde, non-representational art—a factor which perhaps helps to explain the fact that the presence of structuralism was felt most strongly in France and, even earlier, in pre- and post-revolutionary Russia. There were, however, marked differences between the French, predominantly Saussurean and 'scientistic', and the Russian, originally 'literary' and more self-consciously avant-garde, variants of structuralism. It is usual to distinguish the latter by using the term 'Formalism', which may be taken to refer both to a method and to a literary-political movement.

Russian Formalism was originally conceived in fierce reaction to the focus upon imagery, and the frequently religious overtones, of 'Symbolist' poetics; in contradistinction, the Formalists insisted that notions such as 'poetic force' could be explained only in terms of particular uses of language, in particular the employment of 'devices'. The Formalists also argued, against naturalist and realist aesthetics, that the work of art should be judged not by its degree of likeness to reality, but in terms of its transformation of this reality through devices. In this, the Formalists joined forces with the Futurist avant-garde, as represented for example by the poet Myakovsky, who wrote that 'Art is not a copy of nature, but the determination to distort nature in accordance with its reflections in the individual consciousness.'[12] Formalism, then, came to be identified with a tendency to view artistic and literary works from the point of view of device and technique, divorced from considerations of both representational content and socio-historical context. It is hardly surprising, therefore, that Formalism was regarded by the proponents of 'Socialist Realism' as a thoroughly bourgeois deviation; nor that to be accused of it could mean, in the Stalin era, deportation to the labour camps.

Before its suppression, however, the Formalist school developed a number of theoretical positions which distinguished it from Saussurean structuralism. The best known of these concerned the role of the formal device in defamiliarizing ('making strange') and rendering perceptible the linguistic material of the poetic work; a concept which was later rendered by the Prague structuralists as 'actualization' or 'foregrounding'. It was this idea that underlay the proposal by the Formalists of the 'primacy of sound over meaning' in poetry, which was seen as the distinguishing feature of the poetic, as opposed to referential, function of language; and would be developed into a full-blown functional-structuralism by Jakobson and his colleagues in the Prague Linguistic Circle.

In contrast, Saussurean structuralism was a predominantly classificatory enterprise in which meaning was considered to be a property of a *code* shared by speakers of a language; by extension, the analysis of other semiotic systems also consists in the unravelling of the systematics of the code, whether this be myth, art, ideology or ritual. The fundamental methodological tool of Saussurean semiotics (or semiology) is the 'binary opposition', since the contrastive identity of any given code-unit is ultimately reducible to its place in the set of binary oppositions characterising the entire code system. From this theoretical perspective, culture itself is viewed as a 'code' or set of codes transmitted through the generations, and the process of acculturation or socialization is viewed, effectively, as a task of learning of linguistic and other codes. It is easy to see how such a perspective can give rise to a view of cultural and individual world-views as being completely determined by the pre-existent code. Cultural and linguistic determinism and relativism has in fact been a persistent feature of structuralist theory, from Whorf and Sapir to the present day; though it has also been in a constant tension with the universalizing impulse to uncover the 'structure of structures' from which other culturally specific codes are derived, and which has characterized the work of, for example, Lévi-Strauss and Piaget.

The structuralist reduction of representation to signification, and of interpretation to a process of 'decoding', has also been associated in Saussurean structuralism with a general neglect of language uses and functions,[13] and with the organization of linguistic and semiotic material at a level 'higher' than that of the individual sentence or utterance. Not only is language no longer rooted in a naïve theory of representation, but representation is banished altogether from language and discourse. Thus, in Saussurean-inspired structuralist theories, and in some contemporary 'post-structuralist' approaches, the Kantian notion of 'discursive representations' is replaced by reduplicating syntagmatic 'chains' of signifiers, and communicative acts become mere effects of the endless circular movements of a galaxy of signs. Within this theoretical perspective (as the Prague School theorists realized), diachronic change itself becomes, for each sign-system, both arbitrary and self-enclosed; and the nature of communication as exchange is emphasized at the expense of interpretation, information and mediation.

Chomsky and generative linguistics

Saussure's structuralism had as its goal the *description* of the language system (*la langue*) which he took to underlie the actual behaviour of speakers and hearers. This description should therefore be *consistent* with such behaviour (*la parole*), for only through the study of behaviour could the nature of the system be elucidated. A more ambitious goal for linguistic theory was formulated by Noam Chomsky in his now-famous book *Syntactic Structures* (Chomsky, 1957): that is, to *explain* linguistic behavior—*performance*—by the construction of a grammar capable of *generating* the combinatorial structures constituting all the possible sentences of a given language. Such a grammar—and in particular the Transformational Generative Grammar (TGG) proposed by Chomsky—will then constitute a description, not merely of the facts of language, but of the underlying linguistic *competence* of the native speaker–hearer.

Chomsky's work has not only revolutionized linguistic theory, but has also had a profound effect upon language acquisition theories (Wexler, 1982), as well as serving as a paradigm model for theory construction in other areas of cognitive science. Chomskyan linguistics[14] is based upon a distinction between underlying competence and actual performance which, despite its apparent similarities with Saussure's langue–parole distinction, is very different in kind. Competence, as understood by Chomsky, is intended as a description of the actual (though tacit) knowledge of the speaking subject; it is thus an individual psychological—and, ultimately, psychobiological—notion, rather than a social psychological one pertaining to societies and communities, as was Saussure's 'langue'. In this respect, Chomskyan theory falls squarely within the tradition of Cartesian rationalism, in which questions of knowledge are cast in terms of individual subjects, endowed with biologically-inbuilt, but universally applicable, innate capacities (Chomsky, 1968).

Such capacities are defined, as is the Saussuran 'langue', in terms of formal properties. Unlike in Saussurean structuralism, however, the basis of the analysis is neither the morphological minimal unit of contrast, nor its semantic analogue, the unit of sense as 'value'. Rather it is the sentence (more specifically, the infinite, exhaustive and exclusive set of 'all and only the grammatical sentences of a language').

The Chomskyan 'revolution' in linguistics, it should be noted, goes further than merely reworking at a higher and more rigorous level the descriptive enterprise of Saussurean linguistics, even though it shares with the latter a focus on form in isolation from function and communication. For Saussurean linguistics shared, at least in residual form, in the concept of the 'signified', the orientation towards *interpretation* characterizing the entire semiotic tradition. This is not the case with Chomskyan linguistics, and cognate theories, in which interpretation and understanding are considered to be manifest in production (performance). The most obvious example of the way in which this orientation can be said to typify the dominant trends in contemporary cognitive science is the well-known Turing test. This proposes that the criterion for determining whether a given device, or automaton, constitutes a valid theory (simulation) of a given grammar (cognitive process) consists in whether its symbolic output, in interaction with a human speaker (interactant), is distinguishable or not by that interactant from that which would be produced by another human interactant.

Since the question of rules, algorithms and representations is explored elsewhere in this book, I shall not pursue it here, except to note that this feature of Chomsky's theory, amongst others, suggests that it should not really be considered to be a semiotic theory in the same way as the other ones discussed in this chapter. On the other hand, it is also possible to argue, as I do below, that the Chomskyan approach, and by extension the majority of theories in Artificial Intelligence, constitute a logical terminus in the long process whereby representation has progressively been reduced to signification, and mind and mental process to the manipulation of formal (syntactically interpreted) symbolic entities.

Social life as semiosis: Mead, Bakhtin, Barthes

The work of Peirce, Frege and Saussure laid the ground for the subsequent study of language in the disciplines of linguistics, philosophy and psychology, as well as establishing the foundations of the science of semiotics. In this section, I try to characterize the ideas of three further thinkers, whose influence has perhaps not been so uniformly pervasive as those discussed so far, but whose work may be considered as representative of the kind of concerns which have guided many semiotic approaches to the role of symbolization in social life, and may also furnish some insights into the limits of structuralist theories.

George Herbert Mead (1863–1931) stands in direct line of descent from Peirce, as an adherent of the pragmatist philosophy also espoused by James and Dewey. He can equally well be placed, however, in the hermeneutic tradition; within which his work represents one of the earliest attempts to formulate a 'social constructivist' psychology and philosophy. Mead's influence has been widespread in social psychology and microsociology, but of particular interest in the present context is that Mead was centrally interested in the problem of how children come to participate in the world of symbolically mediated social action and interaction. For Mead, language (and symbolization) is not merely a reflection of the external world, but is constitutive both of the social world and, crucially, of the self in that world. 'Meaning', for Mead, resides in the social process as a totality, and it is by virtue of symbolization that intelligent human action, and the 'selves' that originate intelligent action, are possible. Symbolization and the human social order are thus mutually interdependent, since meanings both organize experience, and depend upon the social process for their signification:

> a universe of discourse is always implied as the context in terms of which, or as the field within which significant gestures or symbols do in fact have significance. This universe of discourse is constituted by a group of individuals carrying on and participating in a common social process of experience and behaviour ... a universe of discourse is simply a system of common or social meanings (Mead, 1934: 89).

Thus far, Mead's emphasis on the 'universe of discourse' which imparts significance to signs may be seen as an attempt, within the Peircean tradition, to root semiotic processes in the general structures governing the exchange of signs—an attempt which recalls Saussure's theorization of the relations between langue and parole. Mead, however, was not really concerned with meaning in the Saussurean sense of 'value', and was probably not familiar with Saussurean linguistics. Meaning, according to Mead, is intimately connected with action, inasmuch as actions are governed or 'controlled' by meanings which are inherent in the properties of the social environment. This environment is social both in that it includes the responses of others to the acts of the self, and in that the objects towards which acts are directed are meaningful within the matrix of the social process. In the beginning, the control of acts by meanings is effected by the actual responses of the other:

> Meaning is thus a development of something objectively there as a relation between certain phases of the social act; it is not a psychical addition to the act and it is not an "idea" as traditionally conceived. A gesture by one organism, the resultant of the social act in which the gesture is an early phase, and the response of another organism to the gesture, are the relata in a triple or threefold

relationship of gesture to first organism, of gesture to second organism, and of gesture to subsequent phases of the given social act; and this threefold relationship constitutes the matrix within which meaning arises, or which develops into the field of meaning ... objects are constituted in terms of meanings within the social process of experience and behavior through the mutual adjustment to one another of the responses or actions of the various individual organisms involved in that process, an adjustment made possible by means of a communication which takes the form of conversation of gestures in the early evolutionary stages of that process, and of language in its later stages (Mead, 1934: 76–77).

The Peircean flavour of this quotation is, of course, not accidental; but what is most important about Mead's approach is that he suggests that 'meaning' in the widest sense, is reducible neither to the paradigmatic structure of the field of signifiers, nor to the 'signified' in the traditional sense of 'idea' or 'concept'. Rather, meaning is constituted by the relationship between a communicative act and the social and interpersonal context within which the act takes place. Meaning, to use the terminology of hermeneutics, is intersubjectively constructed; and so, along with it, is the Self. Mead, in effect, attempted to reconcile the romantic-expressivist with the phenomenological tendencies of post-Kantian philosophy, through the elaboration of a notion of meaning as action in social context. He was thus one of the originators of the linguistic or semiotic subdiscipline known as 'pragmatics'.[15]

Mead, like many of his contemporaries, was also profoundly influenced by evolutionary theory, and his account of symbolization attempted to integrate developmental and evolutionary considerations into semiotic theory (in this respect, he was also influenced by the work of J. M. Baldwin: see Russell, 1978; and Chapter 4). The critical evolutionary and developmental achievement of human beings, according to Mead, lies in the use of 'significant symbols' as vehicles for both social exchange and psychological processes. The earliest manifestation of the significant symbol he held to be the 'vocal gesture', which signifies aspects of the 'implicit' meanings of social situations in such a way as to 'lift' them into consciousness, thus constituting the self and reflective intelligence. Many commentators have noted the similarity between this view of Mead's, and the role assigned by Vygotsky (see Chapter 3) to the internalization of speech in the ontogenesis of socialized reasoning (Lock, 1978).

Mead's social-pragmatic account of the nature and development of symbolization is also comparable in many respects to that offered by his Russian contemporary, Mikhail M. Bakhtin (1895–1975).[16] Compare, for example, the following:

> Consciousness or experience as thus explained or accounted for in terms of the social process cannot, however, be located in the brain ... it belongs to, or is a characteristic of the environment in which we find ourselves (Mead, 1934: 112).
>
> Social psychology in fact is not located anywhere within (in the "souls" of communicating subjects) but entirely and completely *without*—in the word, the gesture, the act ... that which has been termed "social psychology" ... is, in its actual, material existence, *verbal interaction* (Volosinov, 1973 [1929]: 19).

Like Mead, Bakhtin held social interaction and process to both constitute, and be governed by, the use and exchange of signs. Further, the sign was seen by Bakhtin, as it was by Mead, to be the vehicle of subjective consciousness and experience:

> signs can arise only on *interindividual territory*. It is territory that cannot be called "natural" in the direct sense of the word ... It is essential that the two individuals be *organised socially*, that they compose a group (a social unit); only then can the medium of signs take shape between them ... individual consciousness is not the architect of the ideological superstructure, but only a tenant lodging in the social edifice of ideological signs ... the reality of the inner psyche is the same reality as that of the sign. Outside the material of signs there is no psyche ... the subjective psyche is to be localised somewhere between the organism and the outside world, on the borderline separating these spheres of reality ... but the encounter is not a physical one: the organism and the outside world meet here in the sign (Volosinov, 1973: 12, 13, 26).

It would be wrong, however, to see Bakhtin as primarily a psychologist of language, as his work ranged extraordinarily widely over philosophy, linguistics, and literary and semiotic theory. Bakhtin's writings were informed by dialectical and historical materialist philosophy, but he was perhaps the first Marxist to acknowledge the lack of an adequate theory of the subject in orthodox Marxism, and explicitly to locate the material substrate of consciousness in the sign.

Bakhtin was familiar with the principles of structuralist linguistics, as developed by Saussure, and with the related work of the Russian Formalist school. Whilst recognizing the scientific advance constituted by the structuralist method, Bakhtin took issue with its dualistic distinction between *langue* and *parole*, and with the priority assigned by it to the systemic analysis of the former. Equally, while recognizing the need to re-instate subjectivity into theories of language, Bakhtin was critical of post-Kantian theories, whether phenomenological or romantic-expressivist, which assumed the primacy of the subject over the means of expression. Bakhtin's attempt to synthesize and transcend this opposition between subject and object took as its starting point the *utterance*, or speech act, as a social and communicative event. The actual and dynamic process of linguistic exchange, asserted Bakhtin, can only adequately be understood by recognizing the inherently *dialogic* character of the sign and signification:

The actual reality of language-speech is not the abstract system of linguistic forms, not the isolated monologic utterance, and not the psycho-physiological act of its implementation, but the social event of verbal interaction implemented in an utterance (Volosinov, 1973: 94).

Bakhtin's linguistic analysis thus focussed not upon the form and structure of language, as revealed by the study of sentences; but upon the negotiation and implementation of speech-acts by interlocutors, within extended discourse. From this analysis, Bakhtin attempted to derive a theory of speech and behavioural 'genres', and of the articulation of the speaking subject within discourse structures characteristic of such genres. Such discursive genres, he argued, must be understood with reference to 'the objective conditions of social-verbal interaction'; that is to say, overall social relations as crystallized in 'behavioural ideologies'.

Paradoxically perhaps, given Bakhtin's indictment of the secondary role assigned by structuralism to 'parole' and his insistence on the primacy of speech, his concrete research addressed the problem of the production of speech and behavioural genres through the analysis of written texts, and particularly in the study of the novel. Whereas in poetry (the favoured domain of Formalist analysis), argued Bakhtin, the writer emphasizes his or her origination of the act of utterance, heightening the direct expression of linguistic design, the voice of the novelistic author is distanced from the linguistic material, which is employed to convey or represent the 'tone' or 'accent' of the depicted speaker or narrative point of view (Todorov, 1984). Thus, in the novel, the inherent 'dialogism' of language achieves a representation in discourse itself, and for that reason the novelistic form was considered by Bakhtin to be representative of the role of utterance and dialogue in discourse generally. In Bakhtin's work, we therefore see, for the first time since Kant's philosophy put an end to classical, rhetorically-oriented semiotics, an attempt to reintegrate representation with discourse, in a theory of utterances and texts.

In contrast to structuralist and formalist linguistics, whose object—signification—consists in 'the relations of a sign to another sign or to other signs (that is, all systematic or linear relations between signs)', Bakhtin proposed a 'translinguistics'[17] of 'the interpretative understanding of non-reiterative utterance ... not tied to the (reiterative) elements of the system of language (that is, to signs), but to other (nonreiterative) texts by particular relations of a dialogical nature' (cited in Todorov, 1984: 50–51).

Bakhtin's theory of the utterance involved attention both to the linguistic and to the extra-linguistic aspects of pragmatic context, which he summarized as follows:

> The boundaries of each concrete utterance, as a unit of verbal communication, are determined by changes in the subjects of the discourse, that is the speakers. Every utterance has a specific interior completion. An utterance does not merely refer to its object, as a proposition does, but it *expresses* its subject in addition: the units of language, in themselves, are not expressive [here Bakhtin is implicitly criticising Humboldt and the other romantic-expressivist linguists]. In oral discourse, a specific, *expressive* intonation marks this dimension of the utterance. The utterance enters in relation to past utterances which had the same object, and with those of the future, which it foresees as answers. Finally the utterance is always addressed to someone ... The extraverbal context of the utterance is composed of three aspects: (1) the spatial horizon common to the interlocutors (the unity of the visible: the room, the window, etc.); (2) knowledge and understanding of the situation, also common to both; (3) their common evaluation of the situation (cited in Todorov, 1984: 42 & 53).

These remarks are remarkably prescient, anticipating many of the present-day concerns of text-linguistic and pragmatic theories (Beaugrande and Dressler, 1981; Eco, 1976; Eco 1979; Sperber and Wilson, 1986) in their concern for the 'background' of utterances in context and co-text. Other statements of Bakhtin have an equally contemporary flavour, such as his proposals that 'Discourse is in some way a "scenario" of a certain event', and yet others anticipate the theories of the Prague School, as with his suggestion that discourse is organized in terms of two groups of functions, the 'elective' (or selective) and the 'compositional'. Bakhtin's most essential and lasting contribution, however, lay in the way that he identified and attempted to circumvent the essential weakness of the (structuralist) semiotic project, almost at the moment of its inception. 'Semiotics', he wrote, 'prefers to deal with the transmission of a ready-made message by means of a ready-made code, whereas, in living speech, messages are, strictly speaking, created for the first time in the process of transmission, and ultimately there is no code' (cited in Todorov, 1984: 56).

Bakhtin's critique of structuralism was unknown for many years outside the Soviet Union, and was at best barely tolerated by the guardians of Socialist Realist orthodoxy within it. Its rediscovery in France in the 1960s[18] coincided with, and contributed to, the re-evaluation of structuralist theories which eventuated in 'post-structuralism'. I shall not attempt the impossible and self-defeating task of defining post-structuralism, and its twin, post-modernism (see Harland, 1987). Instead, I shall adopt the simpler solution of indicating a number of relevant themes with reference to the evolution of the semiotic theories of Roland Barthes (1915–1980).[19]

Barthes' early work (e.g. Barthes, 1957) situated itself firmly within the Saussurean tradition, in attempting to analyse non-linguistic sign systems, such as photography, or fashion, in terms of internal structural oppositions understood with reference to underlying systems of 'ideology as myth'. Barthes came to maintain, however, that such signifying systems themselves depend crucially upon language, inasmuch as language both mediates their representational functioning, and apparently offers an analytical discourse—a 'metalanguage'—capable of effecting the 'unmasking' of this (ideological) function:

> It is true that objects, images and patterns of behaviour can signify, and do so on a large scale, but never autonomously; every semiological system has its linguistic admixture. Where there is a visual substance, for example, the meaning is confirmed by being duplicated in a linguistic message (which happens in the case of the cinema, advertising, comic strips, press photography etc.) so that at least a part of the iconic message is, in terms of structural relationship, either redundant or taken up by the linguistic system. As for collections of objects (clothes, food), they enjoy the status of systems only insofar as they pass through the relay of language ... we are, much more than in former times, and in spite of the spread of pictorial illustration, a civilization of the written word (Barthes, 1984 [1964]: 78).

At the outset, then, Barthes' project was the familiar one of 'de-naturalizing' representation, demonstrating that the most apparently spontaneous and 'realistic' image or text is 'read' through a grid of linguistic signifiers: signs refer to signs, and not to a pre-given (pre-signified) 'untexted nature' (Silverman and Torode, 1980). This de-naturalization of representation, as usually in structuralist-inspired theories, took the form of a reduction to signification. Barthes, however, took the process one step further by suggesting that, not only are other sign systems structured *like* language, they are structured *through* language. Thus, reversing Saussure's definition of semiotics, he wrote: 'linguistics is not a part of the general science of signs, even a privileged part, it is semiology which is a part of linguistics: to be precise, it is that part covering the *great signifying unities* of discourse' (Barthes, 1984: 79).

The significance of this reformulation consists in the way that the traditional relation between 'text' and 'context' is inverted; it is text, according to Barthes, which constitutes the signification of non-textual elements, rather than, as traditionally conceived, the non-textual elements providing a referential grounding and context for the textual ones.

Barthes' next crucial move, marking a decisive break with the Saussurean tradition, came when he extended the 'dialogic' premise of Bakhtin's analysis of utterances to the production of texts. Texts, maintain Barthes, do not arise *de novo*, fully-armed from the brow of the singular author; nor do they merely chart an unproblematic reality, for the 'semblance of reality'—the 'vraisemable'—is itself produced as an effect of signifying processes. Rather, texts can be understood only in terms of their *intertextuality*, as summarized by Kristeva (1970: 69): 'So as to study the structuration of the text as transformation, we shall picture it as a textual dialogue, or better, as an intertextuality.'

Intertextuality is, according to Barthes, a *productive*, and not merely an interpretive, relation; indeed, an effect of the concept of intextuality is to render problematic any purported interpretive meta-language, for the ideal of the meta-language turns out to be precisely the (idealistic) illusion of the transcendental text of nature:

The text is plural. Which is not simply to say that it has several meanings but that it accomplishes the very plural of meaning: an irreducible (and not merely an acceptable) plural. The text is not a co-existence of meanings but a passage, an overcrossing; thus it answers not to an interpretation, even a liberal one, but to an explosion, a dissemination. The plural of the text depends, that is, not on the ambiguity of its contents but on what might be called the *stereographic plurality* of its weave of signifiers (Barthes, 1977: 159).

Thus, other texts are not merely the disambiguating 'context' of the text, but the invoked and evoked, actual and virtual, alterities and 'différances' (Derrida, 1982) which are set in motion through the 'play' of signifiers; through dialogue, parody, contradiction and contestation. Barthes' dialectical method therefore shifts the ground of semiotic analysis away from its initial preoccupation with 'signification'—that is, signifier–signified relations; towards 'signifiance'—that is, the productive work of signifiers in the textual and discursive practices within which 'readings' (and their subjects) are positioned and produced as multiple, and possibly contradictory, sites.

From this perspective, the monologic and univocal 'referring expression' of linguistic analytic philosophy in the tradition of Frege; the 'unity of the sign' of Saussurean linguistics and semiotics; and the individual subject, (the *cogito* of Descartes, the *transcendental ego* of Husserl and the *competence* of Chomsky) are all equally problematized by Barthes, and by other post-structuralist critiques. Not only is representation reduced to signification, but signification becomes a process in which the 'signified' is *produced*, rather than *referred to*, in discourse.

It is instructive to compare this theoretical position with that of Chomsky. Both theories, in different ways, constitute simultaneously logical termini of the semiotic project, and in effect negations of that project. In both of them, the original orientation towards interpretation, judgement and reference is displaced by an orientation to autonomous sign-production processes. In both of them, too, all notions of representation as resemblance or mimesis seem finally to have been cleansed from the theory: in Chomsky, through the formalization of algorithmic relations between signifiers, in Barthes through the positing of reality, object and subject in a signifying order whose origin and reference is text and discourse. Finally, Barthes and Chomsky mirror each other in the one's *removal* of the subject from the privileged centre of the ceaseless re-orderings of signification, and the other's *elimination* of all aspects of subjectivity other than a purified (non-referential) predicative act, the originating locus of linguistic competence. Communication, in the sense of information, is marginalized by both theorists—as a 'filter' of psycholinguistic performance parameters in the case of Chomsky, and as the narrative 'vraisembable' of the realm of the imaginary, in the case of Barthes.

In a curious way, it seems then that the 'death of the subject' proclaimed in post-structuralist theory finds its analogon in the reduction, in contemporary neo-rationalism, of the Cartesian 'I' to the mechanistic combinatorials of a syntax of notation-without-denotation. I shall return to the neo-rationalist paradigm in Chapter 4; but, in closing this chapter, it is worth noting that in the work of both Barthes and Kristeva the 'deconstruction' of both subjectivity and representation is, as it were, a moment in the struggle for their reconstruction—a goal which distinguishes their work from that of, for example, Derrida, whose metaphysics of pure negation has been criticised on precisely those grounds by Kristeva (1986).

In Barthes' critical practice (which was not, it must be said, always consistent with his theoretical statements—perhaps thereby gaining rather than losing in percipience) he repeatedly addressed issues arising from the comparison of text and image. In 'The Photographic Message' he wrote that 'In front of a photograph, the feeling of "denotation", or, if one prefers, of analogical plenitude, is so great that the description of a photograph is literally impossible; *to describe* consists precisely in joining to the denoted message a relay or second-order message, derived from a code which is that of language, and constituting in relation to the photographic analogue, however much care one takes to be exact, a connotation: to describe is thus not simply to be imprecise or incomplete, it is to change structures, to signify something different from what is shown' (Barthes, 1983: 197–198).

It is this notion of the 'analogical plenitude' of representation, the *resistance* of the signified to proposition and predication, which underlies both Barthes' emphasis on the centrality, in signification, of language as a 'relay' or structure of mediation; and his attempt to understand the nature of texts as multivalent pluralities, which through their intertextual signifying play precisely recreate the 'plenitude' of representation, in that 'other structure' which is language.

Notes

1. The concept denoted by the German word *Darstellung*, which implies both that which is represented, and its mode of presenting itself 'for us', perhaps best conveys the sense in which 'representation' is used here.
2. Certainly, neither Aristotle nor the Stoics believed that ideas are signs *in the same way* that a symptom is a sign of a disease, since Aristotle's mimetic theory of representation, in aesthetics as much as in logic, postulates that what is represented is not the individual object but its class or species. Quotation adapted from Derrida, 1982: 75. Original citation W. D. Ross (ed.) *The Works of Aristotle*, vol. 1, transl. E. M. Edghill, Oxford, Clarendon Press, 1928, 16a.
3. John Locke, *An Essay concerning Human Understanding* (1690), bk. III, ch. 2, sect 2).
4. René Descartes, *Discourse on the Method of Rightly Conducting the Reason*, (1637, pt. 5, transl. E. Haldane and G. Ross, Cambridge University Press, 1911, p. 116).
5. Here and elsewhere in the text, the date of original publication of a work, in square brackets, follows the first citation of the work in translation, collection and/or later or standard edition.
6. One way of looking at this is to emphasize that Kant's philosophy was a systematic attempt to save the notion of representation from Berkeley's critique, by subordinating it to judgement. Kant regarded Berkeley's effective elimination of representation—his espousal of 'presentationism' in contrast to Locke's representationism—as a 'scandal', because of its solipsistic implications. By yet another irony, we find that in the contemporary philosophy of mind it is presentationist theories of 'direct perception' which emphasise objectivity, while neo-rationalist representationalism has taken to solipsism; see ch. 4.
7. Kant does not use the term adequation, but if my reading of him is correct the core of his 'revolution' consisted precisely in the replacement of the notion of representation as resemblance with that of representation as adequation; or, as Putnam (1981: 64) puts it, truth as 'ultimate goodness of fit'.
8. Karl Marx, *Capital*, vol. 3, p. 187; Moscow edition, 1971. The distinction between 'essence' and 'appearance' can be traced back at least as far as Aristotle, who wrote: 'there is epistemic knowledge of a thing only when we know its essence' (*Metaphysics*, VII); and has more recently been expressed by Fodor and Pylyshyn (1981: 149) thus: 'You would not really expect the properties in virtue of which objects satisfy laws to be coextensive, in the general case, with those which are phenomenologically accessible. If such a general coextension held, doing science would be a lot easier than it has turned out to be.'
9. More accurately, Frege transformed the transcendental problematic of Kant into a formal and analytic problematic; while, as Apel (1980: 78), points out, Peirce undertook 'a semiotical transformation of Kant's transcendental logic'.

10. Saussure used the expression 'value' for what is usually referred to as 'sense'.
11. See Merquior, 1986, for a recent thorough and readable review.
12. Cited in Erlich, 1969: 46. Erlich's study of Russian Formalism remains the most complete to date, though it lacks extended consideration of the work of the Bakhtin circle. For a more recent study, see Steiner, 1984.
13. The integration of the embryonically functional orientation of Russian Formalism with the contrastive method of Saussurean structuralism was, in fact, the particular theoretical project of the Prague School.
14. Linguistically knowledgeable readers will be aware that the 'standard', 'extended standard' and 'revised extended standard' theories have now given way to the theory of government and binding (Chomsky, 1981). As I indicate in ch. 4, although I am critical of the implications for cognitive science which Chomsky and others have thought to draw from generative linguistics, I make no attempt to evaluate these approaches *qua* theoretical linguistics.
15. Morris (1938) defined syntax as the study of 'the formal relation of signs to one another', semantics as the study of 'the relations of signs to the objects to which signs are applicable' and pragmatics as the study of 'the relation of signs to interpreters'. More recent attempts to define linguistic pragmatics have suggested that it is the study of all aspects of meaning that do not fall under semantics; that it is the study of all aspects of meaning not accounted for by truth conditions on sentences; or the study of all aspects of context relevant to grammar. See Levinson, 1983.
16. As is now widely known, many of Bakhtin's writings were published under the names of co-authors; in particular, P. M. Medvedev and V. N. Volosinov. The circumstances of Bakhtin's work are as fascinating as the work itself (see, for example, Clark and Holquist, 1984; Todorov, 1984), but lie outside my present scope.
17. This is Todorov's term for what Bakhtin called 'metalinguistics'; 'metapragmatics' would be an equally appropriate rendering.
18. The entry of Bakhtin into French intellectual life predated his 'rediscovery' in the English-speaking world by almost two decades (though see the review of Volosinov (1929) by Henriques and Sinha, 1974). This no doubt partly accounts for the fact that, whereas in France Bakhtin's work was seen as a major source for the development of post-structuralist theories, it is currently appropriated by many Anglo-Saxon authors as a source from which to develop a critique of post-structuralism. This has led to an abundance of ironies of a sort that Bakhtin would no doubt have relished: it is curious, for example, that his work, which fell foul of the Stalinist 'Battle against Formalism', should have been enlised in a latter-day attempt to re-enact the campaign: see Williams, 1986.
19. Because of the considerable, and mutual, influences exerted upon each other by Barthes and his colleague Julia Kristeva (e.g. Kristeva, 1986), who, together with her Bulgarian compatriot Tzetvan Todorov, introduced Bakhtin to French-speaking readers, it is inevitable that much of the following refers directly or indirectly to her work, as well as that of Barthes.

2 The Dialectics of Representation

We speak of a picture representing a subject, a lawyer his client, a sample the species of which it is a part, or again of a symbol representing the powerful but perhaps obscure realities to which it holds the key ... there is therefore in all genuine representation a dialectic between its appearance and the content it is to convey.

Richard Bernheimer (1961: 25–26)

So you should simply make the instant
Stand out, without in the process hiding
What you are making it stand out from ... permitting the spectator
To experience this Now on many levels, coming from
Previously and
Merging into Afterwards, also having much else now
Alongside it. He is sitting not only
In your theatre but also
In the world.

Bertolt Brecht

The conditions on representation

To represent something—a scene, an event, an object, an interest—is to cause something else to stand for it, in such a way that both the relationship of 'standing for', and that which is intended to be represented, can be recognized.

The above is a definition of the conditions on what may be called the *canonical case* of representation; that is, in which a representative subject recognizes the representational status of a representation, and refers it appropriately to that which it is intended to represent. Even this apparently loose definition constitutes something of an idealization, for it supposes successful or 'transparent' representation. It is analogous to Grice's Co-Operative Principle in pragmatics (Grice, 1975), which enjoins speakers to be informative, truthful, relevant and clear; and hearers to work on the assumption that speakers are so behaving. This is, evidently, an idealization of, and abstraction from, the not infrequently deceitful, irrelevant and mystificatory motives in play in actual communicative acts. Nonetheless, it provides a canon, or rule, against which the forms and practices of what Habermas (1970) calls distorted communication may be measured.

Grice's Co-Operative Principle, then, is an example of what I shall call a *canonical* rule or regularity[1]; a type of rule whose role in communication and cognitive representation, and the acquisition and development thereof, I shall stress throughout this book. Canonical rules, unlike laws and propositions, are not defined truth-functionally; and to that extent they possess a degree of conventionality (but *not* arbitrariness). Canonical rules, I shall suggest, are best seen as having a status midway between, or at the interface of, a number of distinctions which underly many areas of social scientific theory. Thus, canonical rules cannot be seen as strictly definitional of behavioural and communicative acts, since instances in which they are violated do not disconfirm them, but shed light upon the processes underlying them.[2] Neither, however, are canonical rules merely stipulative, or normative in a statistical or heuristic sense; rather, they provide a fundamental basis for the intelligibility of social behaviour both within and across linguistic communities. As I shall suggest in Chapter 5, at least some types of canonical rules can also be seen as interfacing or mediating the domains of the 'natural' and the 'socio-cultural'. For the time being, however, the meaning of 'canonical' can be glossed as 'socio-culturally standard'.[3]

Applying this notion to the definition offered above of representation, let us examine some instances in which the conditions for a canonical case of representation are not met. The first can be illustrated by Figure 2.1.

Although the figure appears at first sight to be non-representational—that is, unrecognizable as a respresentation—the information that is given by its title[4] enables the observer to infer, and thus to 'see' that which it represents, by imposing a *recognitory schema* upon the representation. Oatley (1978) discusses similar examples to illustrate the inferential processes underlying the perception of what is 'represented' by incomplete or degenerate sensory information; and such phenomena played an important part in the accounts proposed by, amongst others, the Gestalt psychologists and the 'New Look' perception theorists of the 1950s.

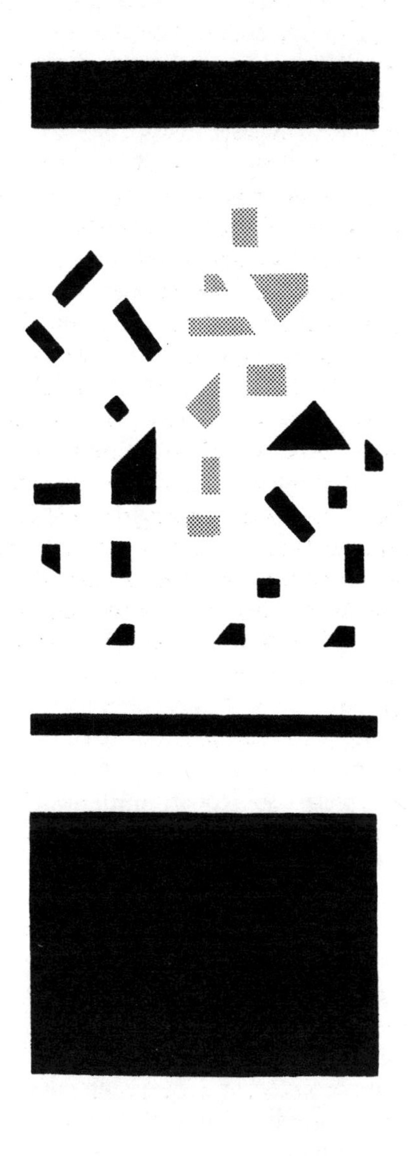

Figure 2.1: A representation?

For present purposes, let us note that Figure 2.1 departs from the canonical case of representation primarily by virtue of the difficulty experienced by the subject in recognizing that which is represented, and only secondarily by virtue of the misrecognition of the representative status of the figure (though this is no doubt an intended effect). In an historical context, these two aspects of the canonical representative schema have not always been distinguished. When the Impressionist movement in painting initiated modernism, by transforming the canons governing the recognition of that which is represented, critics accused them of violating the *entire* canon of representation by producing 'meaningless' figures devoid of representational status: that is, not merely unrecognizable pictures, but not pictures at all (a step which was not in fact taken until the advent of 'abstract' or 'nonrepresentational' art). These critics failed to realize that what defines the representational status of the representation is the *recognized intent* to represent, and not the form of the representation—its verisimilitude or 'realism'.

Let us turn now to a second example. In an earlier period of representative realism in painting, artists would sometimes produce pictures or figures which merged so imperceptibly into their immediate visual context that, to the observer, they were indistinguishable from it. Such *trompe-l'oeil* figures, usually miniatures,[5] were intended to deceive the eye by concealing from it their *status* as representations. To use Gregory Bateson's terminology, they were so constructed as to occlude or misrepresent within themselves the principle that 'the map is not the territory, and the name is not the thing named' (Bateson, 1973: 37). *Trompe-l'oeil* pictures are second-order representations, in which the object of representation is representation itself, conceived as verisimilitude. Their effect is dependent upon the mastery of the canons of first-order representation, which consisted, for post-Renaissance painters, of the translation onto canvas of a 'code of nature' (optics, perspective and so forth). Not only do these pictures deceive, but they are intended to deceive, and are so constructed as to produce a deception effect.

The principal issue highlighted by these examples concerns the centrality of *intention* and of *recognition* in representation. The definition I have offered of representation allows various different types of misrecognition to occur, any of which can lead to a violation of the conditions on the canonical case of representation.

First, and as illustrated above, the subject may fail to recognize that what he or she is encountering *is* a representation—the subject may misperceive the representation as a non-representational design, pattern or structure (such as might be found on a rug, or on the wall in the foyer of a building owned by a large corporation). In such a case, the representation is not referred to a represented object but is perceived as a *motif*. A motif, in this sense, is a vague or contentless (though not, by definition, formless) 'pararepresentation' which, though it may serve an aesthetic or decorative purpose, does not seek to 'stand for' something outside itself. Motifs may be seen as visual equivalents of what, in language, Malinowski (1930) called 'phatic communion', which 'serves to establish bonds of personal union between people brought together by the mere need of companionship and does not serve any purpose of communicating ideas'.[6] In the case of Fig. 2.1, misrecognition occurs because the representation is misrecognized as a motif (as a result of the difficulty experienced in organizing the design *as a representation*).

Similarly, the representational import of a communication (its *informational* content) may be unclear, and the communication may fail, not because communicative intent is unrecognized, but because the listener (or interpreter) may fail to attribute a representational or informational intent to the communication, misperceiving it as phatic communion. We may think, for example, of Stevie Smith's poem 'Not Waving But Drowning', in which a call for help (a representation of need) is misunderstood as a simple phatic gesture or motif ('Here I am').

Simple motifs, or contentless patterns, do not normally either 'signify' or 'represent'. Nonetheless, the repetition of a motif, throughout a work or across a series of works, may *come to* signify, in at least the limited sense of establishing its own *identity* as a recognizable *object of experience*, despite variations in its presentation. The establishment of recognizable motifs, in this sense, depends upon, and is a function of, human capacities for categorical perception, and not necessarily upon cognitive processes in the sense of conceptual understanding and judgement.

However, such recognizable motifs may also eventually serve as true indexical signs, in the sense that they permit the subject to identify either the originator or the bearer of the motif ('that is a typical work of X', 'that is typical of the genre X as a whole', etc.); or as true iconic signs, in the sense that they enable the subject to identify another work or class of works which is copied, parodied, burlesqued or alluded to. In the latter case, we can see how stylistic effect may be achieved through the 'doubling' of representation through signification, in which an initial non-representative object (a motif) becomes a signifier for a representational content (the originating context of the motif). It is this kind of doubling, often involving ironic distancing, that is the focus of Bakhtin's and Barthes' analyses of dialogic and intertextual processes (see Chapter 1). Such doubling effects, whether or not with ironic and citational intent, may also set in motion a process by which motifs become *codified*.

There, is, indeed, a special sub-class of codified motifs, such as logos, flags and theme tunes, whose actual *purpose* is to identify their bearers. Such identificatory motifs—which we might call *emblematic signs*—fall somewhere in between true representations and mere decorative or phatic motifs, since while they do not strictly speaking represent an object separate from themselves, they do permit, through design, the recognition, individuation and classification of the object which bears them. Emblematic signs are therefore analogous to proper names in language; and, like proper names, they may incorporate, even if only in vestigial form, descriptive and identificatory information in their own structure. It is well known, too, that the connotative and affective value invested in emblematic signs can be very great: an insult to the flag, for example, being taken as equivalent to an insult to the person or the nation.

We might suggest, then, that the emotive potency of emblematic signs is a consequence of their function being to represent, for example, a nation's identity, in the same way in which an ambassador represents a nation's interests. On this interpretation, the identificatory function of an emblematic sign is achieved through that sign 'standing for' a larger entity of which its bearer is a part—a relationship which is classically known as *synechdoche*, one of the *metonymic* relationships (see Chapter 5). The emblematic sign, in isolation from its bearer, will then be interpreted as a metaphoric, symbolic sign for the whole of which the bearer is a part, and as such appears to fulfil the conditions on representation which were defined at the beginning of this chapter. Is there, then, a general relationship between representation and signification, such that the former is reducible to a species of the latter?

This has certainly been a common view, and has frequently taken the form of an identification of representation with iconic and/or symbolic signification. In the following Section, I shall argue that, although in cases such as emblematic signs, the mechanisms of representation may well be elucidated through semiotic analysis, the simple identification of representation with any particular semiotic structure is misleading.

Representation and signification

As well as cases, such that illustrated by Fig. 2.1, in which a representation is mistaken for a (simple) motif, there may also be instances in which a motif is mistaken for a representation. Putnam (1981: 1) gives the following example: 'An ant is crawling on a patch of sand. As it crawls, it traces a line in the sand. By pure chance the line that it traces curves and recrosses itself in such a way that it ends up looking like a recognizable caricature of Winston Churchill. Has the ant traced a picture of Winston Churchill, a picture that *depicts* Churchill?' Putnam concludes that the ant has not produced a representation of Churchill, because it has no mental (intentional) apparatus by means of which it can refer to Churchill. In the terms I am using, the ant has produced a motif, and not a representation.

It might be objected, though, that while, from the ant's point of view, the lines in the sand are not intended to, and therefore do not, represent Churchill, nonetheless the *recognition* by a human observer of the similarity between the lines in the sand and Churchill *transforms* the motif into a representation. Supposing, for example, the observer takes a cast or photograph of the pattern in the sand and later puts it on the wall, under the title 'Churchill'. Then, surely, the pattern in the sand (or a copy of it) has acquired representational status. And, if that is so, can't we say that the original pattern in the sand did not represent Churchill for the ant, but did for the human observer?

If this argument were correct, it would mean that 'representation' becomes a function, not of the intentions of the producer, but of the interpretive cognitive structures of the interpreter. We would then, however, be faced with the unwelcome consequence that the design in Fig 2.1 is, in the usual case of being viewed, at one moment (before knowing its title) not a representation, and at another moment (after knowing its title) a representation after all.

The solution to this problem is to distinguish between *representation* and *signification*. On the account I am suggesting, the pattern in the sand can, without difficulty, be said simultaneously *not* to be a representation of Churchill, and yet be capable of being a *signifier* for the interpretant 'Churchill' for some observer. This implies, however, that the observer (or interpreter) has access to some further sign for Churchill, that enables her to *represent* Churchill, both to herself and to others (for example, by pointing to the pattern in the sand and uttering 'That looks like Churchill' to a companion).

By this argument, a copy or cast of the pattern in the sand, when placed on the wall, does indeed function as a representation of Churchill; this representational function is potentiated by the properties of the pattern as a *sign* (in this case an iconic sign), but it is not determined by them—rather, it is determined by the intention to represent on the part of the producer of the sign, who is now the person hanging the copy on the wall, and not the ant. Signification, on the other hand, does not fundamentally depend upon the intentions of a producer—smoke signifies fire for a subject, not by virtue of someone (or God) intentionally producing the signifier 'smoke', but by virtue of the cognitive and inferential processes of the interpreting subject. It was these sub-personal processes that led to the fictional discoverer of the pattern in the sand realizing that the significatory potential of the pattern could qualify it to serve as a representation.

If signification does not determine representation, it is nonetheless necessary for it. There can be no representation—to misquote the American revolutionaries of 1776—without signification, since only through the medium of a sign can a representation attain the materiality and graspable character which enables the intention to represent to be apprehended by another subject. Thus, all representations are simultaneously signs, whereas not all signs are representations. Furthermore, this distinction is not the same as the distinction between iconic (or symbolic) and other signs, although it does indeed, to an extent, map onto it.

As I have already suggested, it is certainly a mistake to underestimate the iconic aspects of symbolic systems, and particularly language. In fact, a sentence or a proposition *represents* a situation just as much as does a picture or diagram.[7] Furthermore, as Langacker (1987) argues, not only is semantic representation irreducible to truth-value, but choices of grammatical structures reflect the adoption of perspectives on, and attitudes to, situations represented (see pp. 52–53). It is also a mistake, however (the very mistake that Kant identified—see Chapter 1) to identify representation with iconicity *per se*. The utterance 'A pound of these, please', made while pointing to a box of tomatoes in a greengrocer's, signifies (by means of a combination of symbolic and indexical signs) a request, and appropriately represents both the object of that request and the intention to make the request,[8] without introducing any iconic material.

It is self-evident that symbolic signification is not in itself a prerequisite for representation, since pictures are representations *par excellence*; but it is less obvious that indexical signs, uncombined with either iconic or symbolic signs, can function as representations, since in general they signify by 'picking out' what they signify, rather than by 'standing for' it. To this extent, the identification of representation with iconic and/or symbolic signification is certainly motivated. Nonetheless, it is not difficult to find examples in which a minimal indexical sign may fulfil the conditions on representation. Suppose, in answer to the question 'What did you do this afternoon?', I respond by pointing out of the window to a newly planted row of cabbages. In this case, although the inferential processes of the person to whom the reply is given are called upon, they are intended to be called upon; the reply is not a motif but a representation.

Thus, although the semiotic principles underlying, for example, linguistic, gestural and pictorial representations differ very considerably, the logic of representation is invariant across the semiotic media. To repeat, the logic of representation involves intentionality, whereas that of signification need not.

It is important to note that the intention underlying representation is a double one. As I emphasized in the definition at the beginning of this chapter, to qualify as a representation a sign must not only be intended to represent something, but must also be intended to be recognizable, in the first place as a representation (not an object or motif), and in the second place as a representation *of* whatever it is specifically intended to represent. This 'double articulation' of representational intent, and the centrality of intentionality in the definition of representation, was expressed by Bernheimer as follows:

> The law of the supremacy of representational intent, as we shall call it, thus follows from another rule ... that likenesses carry the evidence of their function in themselves, and must therefore be read as testimonials to their own purpose ... The inner circle of legitimate representamina is thus surrounded by an outer one of those that fail to live up to minimum logical demands, either because the claims of the rationale are not recognised—as in the case of dummies, models and effigies—or because their marker has failed to make the semblance sufficiently complete. Some of them overshoot the mark and land, as it were, in the field of primary reality, while others do not even manage to attain full status as representamina (Bernheimer, 1961: 141–142).

Sperber and Wilson (1986) have offered an analysis of communicative acts which is similar in some respects to the analysis of representation that I have suggested.[9] They suggest that communication involves:

(a) An Informative intention: to inform the audience of something;

(b) A Communicative intention: to inform the audience of one's informative intention (Sperber and Wilson, 1986: 29).

They later redefine these as follows:

(c) Informative intention: to make manifest or more manifest to the audience a set of assumptions;

(d) Communicative intention: to make it mutually manifest to audience and communicator that the communicator has this informative intention (pp. 59 & 61).

It is not necessary to expand on Sperber and Wilson's terminology to examine the similarities between their analysis and the present one. What I shall suggest is that Sperber and Wilson's analysis is complementary to the definition of the conditions on representation that I have offered, and which for convenience is repeated here in shortened form:

(e) To represent something is to cause something else to stand for it, in such a way that both the relationship of 'standing for', and that which is intended to be represented, can be recognized.

Sperber and Wilson's 'Communicative intention' seems to correspond to the condition on representation that the relationship of 'standing for' should be recognizable—that is, that it should be 'manifest' both to the interpreter and also (*pace* Putnam's ant) to the producer of the representation. Their 'Informative intention' seems, on the other hand, to correspond to the condition on representation that whatever is intended to be represented should also be recognizable, or manifest. Such manifestness, or recognizability, need in neither case be a property of a single sign-unit in isolation. For example, it is frequently unnecessary to render the 'Communicative intention' explicitly and independently of the 'Informative intention', although such formulaic expressions as 'by the way' can serve such a function.

The self-same sign-unit may also subserve both Communicative and Informative intentions—for example, the title of the picture in Fig. 2.1 makes manifest (or more manifest) both the Informative intention (by rendering the figure recognizable), and the Communicative intention (by making it manifest that the figure is intended as a representation). Such inscriptions and 'frames' are important devices, not only in visual representation, but also in linguistic communication, and may be systematically marked—for example, in the Bristol dialect of British English a narrative-initial 'framing' utterance may receive rising intonation. Generally speaking, inscriptions and frames are devices which orient the interpreter towards the activation of a set of relevant 'assumptions', in Sperber and Wilson's terms, against which the principal sign-unit (a picture, a narrative etc.) can be more readily recognized and referred to that which it represents, in cases in which the principal sign-unit cannot easily be read as a 'testimonial to its own purpose'.

It can also be demonstrated, by using one of Sperber and Wilson's own examples, how the distinction between representation and signification can clarify the relationship between language and context. The example is as follows:[10]

Mary wants Peter to mend her broken hair-drier, but does not want to ask him openly. What she does is begin to take her hair-drier to pieces and leave the pieces lying around as if she were in the process of mending it. She does not intend Peter to be taken in by this staging; in fact, if he really believed that she was in the process of mending her hair-dryer herself, he would probably not interfere. She does expect him to be clever enough to work out that this is a staging intended to inform him of the fact that she needs some help with her hair-drier. However, she does not expect him to be clever enough to work out that she expected him to reason along just these lines. Since she is not really asking, if Peter fails to help, it will not really count as a refusal either.

Sperber and Wilson comment: 'Mary does intend Peter to be informed of her need by recognizing her intention to inform him of it. Yet there is an intuitive reluctance to say that Mary *meant* that she wanted Peter's help, or that she was *communicating* with Peter in the sense we are trying to characterise.'

But, of course, it all depends on what we mean by 'mean'. What we can say is that Mary has contrived a situation which does precisely 'mean' that she wants Peter's help, inasmuch as it *signifies* a request. The request is not, however, *represented* as being such—and so Mary can honestly say that she has not *made* or 'meant' a request. In this case, although the signifying structure has been intentionally produced, it has not been produced in such a way that it can be read as a representation of that intention—the signification, and hence the meaning, is real enough, but it is opaque to any reading of it which would allow Mary's intention to be attributed to her.

This again illustrates the proposition that signification encompasses a wider field of phenomena than representation, both in principle and in everyday interactions. Thus, most 'para-linguistic' devices contribute to meaning by way of signification, rather than representation. Facial expression and intonation, for example, may convey (*signify*) the speaker's attitude to what is being represented, without affecting *what* is intended to be recognized as being represented in the utterance; though it should be noted, too, that they may also be used to explicitly *represent* the purpose of the utterance, by enabling the listener to identify whether the speaker is making an assertion, asking a question etc.

Furthermore, what is signified may be outside the deliberate control of the producer. Features of accent and dialect, for example, may affect listener's evaluations of speakers and messages (Giles, 1973). To that extent, they are 'significant' features of language, contributing to 'speaker's meaning for the listener' in a socio-linguistic context. Most semantic (and many pragmatic) theories would not, however, incorporate such features in their accounts, because they do not fall within the purview of either truth-based or intentionality-based views of meaning.

Given this analysis, it would be wrong to identify 'meaning' exclusively with either representation or signification. It is clear not only that natural signifiers (i.e. those which do not depend on human agency) can be interpreted as meaningful, but also that non-representational motifs can also signify without involving any Informative intention beyond the informing of the interpreter of the Communicative intention (as in pure phatic motifs). It would be arbitrary to exclude these phenomena from the realm of meaning, and furthermore would conflict with the intuition that intended, represented meaning is linked in some fundamental way with the 'meanings' in the world in which we perceive and act, as well as communicate. For this reason, I shall suggest that 'meaning' is a general property of sign-systems and sign-usage, and that it embraces both *representational meaning*, defined in terms of the conditions on representation, and *contextual meaning*, which translates as 'all non-representational aspects of a signifying situation'.

In this sense a picture, as well as an utterance in language, can both represent, and signify, and both of these are equally aspects of meaning. The distinction between representation and signification therefore also assists us in clarifying the confusion which surrounds the frequent (and erroneous) dichotomy between 'meaning' and 'representation', as exemplified in the following quotation from Gombrich (1985: 20); 'Only in the discussion of language can we distinguish between statements that have a meaning and strings of words that are devoid of meaning ... A painting of a moonlit landscape does not "mean" a moonlit landscape, it represents one.'

To use Gombrich's example, a representation of a moonlit landscape may, depending upon how it is depicted, signify an attitude of enchantment by natural beauty, or desolation, loneliness and fear. While the signification (or significance) of a visual image may in part be determined by representations which are intended as conventionally symbolic (such as an apple 'standing for' forbidden knowledge), the response evoked by a painting reaches *beyond* representation, whether literal or symbolic, to non-represented (contextual) aspects of meaning, simply by virtue of the fact that a painting is a signifying structure as well as (or even without being) a representation. At least in this limited sense, it is also possible to speak of a 'language' or 'languages' of art, and of artistic effect as being achieved (or read) through signifying structures.

By the same token, any rule-governed use of a signifying system, including natural language, will tend to be 'read' *as* significant, even when there is no apparent representational content. This can be seen in examples such as Lewis Carroll's nonsense poem 'Jabberwocky', or even Chomsky's celebrated 'meaningless' sentence 'colorless green ideas sleep furiously'. The 'contextual meaning' which lends 'significance' to such strings is nothing other than the structure of the signifying system itself.

It is important to add here that 'contextual meaning' is not independent of representational meaning—'context' is not something added to representation which enables the interpreter to 'fill in' what is not represented. Rather, there is a dialectical relationship between representational and contextual meaning, in which each conditions the signification of the other.

Some examples may help to make this clear. First, let us suppose that, in Sperber and Wilson's example, Peter, perceiving the dismembered hair-drier, says 'Would you like me to mend that?', and Mary replies 'Oh thank you Peter, I was hoping you'd do that.' Peter's utterance demonstrates that he has read the signifying situation as Mary hoped he would, and Mary's utterance re-contextualizes the signifying situation by representing it as a request-offer-acceptance sequence. The contextual meaning initially produced (but not represented) by Mary has thus become a part of (mutually—thus successfully) represented meaning, and the politeness routines associated with such a representation may be engaged in, without Mary having to have formulated (represented) an initial request. The different uses of the deictic term 'that' by the two speakers also depend differentially on contextual and represented meaning: Peter's use being disambiguated by extra-linguistic (signifying) context, and Mary's use by Peter's previous utterance in which an offer of help is represented.

A different kind of illustration of the difference between representational and contextual meaning is provided by Shelley's sonnet 'Ozymandias', which is reproduced below:

> I met a traveller from an antique land
> Who said: Two vast and trunkless legs of stone
> Stand in the desert ... Near them, on the sand,
> Half sunk, a shattered visage lies, whose frown,

> And wrinkled lip, and sneer of cold command,
> Tell that its sculptor well those passions read
> Which yet survive, stamped on these lifeless things,
> The hand that mocked them, and the heart that fed:
> And on the pedestal these words appear:
> 'My name is Ozymandias, king of kings:
> Look on my works, ye Mighty, and despair!'
> Nothing beside remains. Round the decay
> Of that colossal wreck, boundless and bare
> The lone and level sands stretch far away.

Clearly, the significance of the lines which Shelley encloses in quotation marks derives *both* from their representational meaning, *and* from the surrounding context (or, in this case, co-text). The poetic effect is not invested singly in either text or co-text, but precisely in the relationship between the two.

Although representational meaning is by definition intentional, the difference between meaning as signification, and meaning as representation, does not correspond to a simple dichotomy between 'intentional' and 'non-intentional'. As the example of Mary's hair-drier shows, signification may be intentional without being representational. Nor is it simply a matter of 'meaning from the producer's point of view' versus 'meaning from the interpreter's point of view'. Rather, representation involves, canonically, a *relationship of identity* between what the producer intends to be represented, and what the interpreter understands the producer to have intended to be represented. It is thus a matter of the *mutual recognition* by producer and interpreter, both of the representational intention of the producer, and of the object under representation. Although such mutuality is definitional of the canonical case of representation, it may equally be seen as a goal of representation (and communication), and as a regulative principle underlying the negotiation of (representational) meaning.[11]

Signification, on the other hand, is more than merely the means by which the identification of what is to be represented is achieved, since it involves all of contextual meaning. Signification is not only a necessary condition for representation, but it also tends continually to defeat the ideal transparency to which representation aspires. Where representation attempts to render an object (situation, scene etc.) in its specificity, even when doing so by means of concepts and categories, signification ceaselessly reduplicates the represented, multiplying meanings and resituating the represented in a network (or vortex) of signifiers. Because representation can proceed only by signification, it is continually subverted, each new appropriation of the sign forcing into existence a specular world of virtual meanings, the skeletons of previous appropriations and the ghosts of choices unmade.

If representation, in Barthes' terms, struggles to reproduce the 'analogical plenitude' of the signified in a manageable, negotiable and mutually-manifest intersubjectivity, signification opens representations to the predations of discourse, and through this aperture re-articulates representation, not just with 'the world under representation', but with the world *as* its discursive representations.

Verisimilitude, truth, reference

The successful representation of something involves, I have argued, the mutual recognition, by the producer and the interpreter of the representation, of that which is intended to be represented. The representation itself may be more or less adequate to its object, but the extent to which the representation facilitates the identification of the object is not linked in any simple fashion to the number of features which are shared by object and representation.

A sketch, or a caricature, may be as readily identifiable as a photograph, and may be evaluated as more 'true' or 'revealing' than a photograph, despite the greater verisimilitude of the latter. Although verisimilitude, or likeness, is implicated in some way in the processes by which we recognize non-discursive representations, such as pictures, the processes underlying visual perception seem not only to be more complex than the notion of a 'copy' would suggest, but also to be susceptible to influence by the discursive context within which the representation occurs, as was illustrated by Fig. 2.1.

Furthermore, in what sense can an utterance in language be said to share a likeness with what it represents? Identification, in language, of what is represented is frequently achieved by means of naming, rather than through descriptive representations—though as we shall see the relationship between naming, description and identification is a complex one. Even assuming that what is named is appropriately named, what underlies the adequacy or inadequacy of a linguistic representation which involves relations *between* named items?

Usually, the notion of 'propositional truth' is assumed to equate to representational adequacy in this sense. So, for example, the expression "Snow is white" is true if and only if snow is white. But, in many cases, there will be several available alternative expressions for representing the same 'facts', all of which bear the same truth value, but whose different degrees of adequacy are pragmatically determined. If a visitor asks me "Have you got a loo?", the alternative answers "Yes" and "It's the second door on the right down that corridor" may both be true replies, but one is clearly more adequate than the other. Again, "Your keys are on the table" is usually a more adequate discursive representation than "The table is under your keys."

Adequacy of representation, and propositional truth, cannot then simply be identified with each other. Suppose, again, in the example of Mary's hair-drier, Peter percipiently starts the interaction by asking Mary 'Does this mean you are asking me to mend your hair-drier?'. Mary can truthfully reply 'No', but the extent to which this would be an adequate representation of the situation is questionable. Adequacy of representation does not simply relate to the truth of isolated propositions, but has to do with the provision of sufficient and relevant information. It is, presumably, for this reason that witnesses in courts of law are required, not merely to speak the truth, but to also to speak the whole truth, and nothing but the truth.

A further problem is that which was alluded to above, namely, that of the relationship between naming, description, and truth. At this point, it is necessary to introduce some futher terminological distinctions with respect to linguistic meaning. Both Frege and Peirce (see Chapter 1) employed the notion of 'reference' more or less co-terminously with that of 'denotation'. As Dummett (1973: 94) puts it, 'the notion of reference, for proper names, thus coincides with that of denotation, as used in the standard semantics'.

The definition, in standard semantics, of denotation is expressed by Lyons (1977: 207) as follows: 'By the denotation of a lexeme ... will be meant the relationship that holds between that lexeme and persons, things, places, properties, processes and activities external to the language system.' Frege's analysis of the distinction between sense and reference was intended to demonstrate, not only that words with different senses may share a common referent, as in the example of the Morning and the Evening stars, but also that the truth-conditions for the reference of a word are determined by its sense: 'Any feature of the meaning of a word which does not affect the reference that it has does not belong to its sense: it in no way follows that two words which have the same reference must have the same sense' (Dummett, 1973: 91).

Frege failed to take into account, however, that pragmatically successful (contextually situated) co-reference may be established even in cases where the truth-conditions for the actual (sense-determined) reference of the constituent terms are violated. Consider, as did Donnellan (1966), the referring expression 'Smith's murderer', used, in the situation where Jones has been wrongfully convicted of murdering Smith, to refer to Jones. Clearly, in such a case, reference (or representation, as defined by the conditions on representation that I have given) can be communicatively successful—that is, referential intersubjectivity may be established—and yet false or untrue. We might say that the *reference* of the term 'Smith's murderer' does not, on this occasion, correspond with its *denotation*. On this basis, Donnellan distinguished between *referential* and *attributive* uses of definite descriptions, the former being usage in order to 'pick out' a particular referent on a given occasion of discourse, and the latter being usage in order to 'designate' or 'denote' that object satisfying the truth-conditions on reference of the definite description, which are determined by the senses of its constituent terms. As Johnson-Laird and Garnham (1980) have noted, this distinction has profound implications for truth-conditional semantic theories.

The argument is simply this. In order for referential intersubjectivity to be achieved (or, *mutatis mutandis*, the conditions on representation to be satisfied), it is not necessary for there to be mutual agreement, or equivalent knowledge, regarding every aspect of the 'meaning' of the terms employed in the referring expression. Indeed, as Putnam (1975) points out, even in cases where reference is *both* successful *and* true, it need not be (and frequently is not) the case that the 'meanings' of the referring expression for speaker and hearer are complete, or correct, or identical. Putnam illustrates his argument with reference to the everyday term 'gold':

> Everyone to whom gold is important for any reason has to *acquire* the word "gold"; but he does not have to acquire the *method of recognizing* if something is or is not gold. He can rely upon a special subclass of speakers. The features that are generally thought to be present in connection with a general name—necessary and sufficient conditions for membership in the extension, ways of recognizing if something is in the extension ("criteria"), etc.—are all present in the linguistic community *considered as a collective body*; but that collective body divides the labour of knowing and employing these various parts of the "meaning" of "gold". This division of linguistic labour rests upon the division of non-linguistic labour (Putnam, 1975: 145).

This example clarifies an extremely important difference between non-discursive representations, such as pictures, and linguistic representations, such as names. The procedures by which a non-discursive representation, loosely based upon some principle of similitude, is recognized as representing that which it is intended to represent, are the *same* as the procedures by which that which is represented is itself recognized. If a subject is presented with some visual figure, such as a painting or a photograph, the recognition of the representation corresponds,[12] cognitively, with the recogition of that which is represented.

This holds even for a figure such as a map, whose essential purpose in fact is to enable the reader to orient him or herself within the territory mapped. In an important respect then, we must modify the quotation from Gregory Bateson earlier in this chapter: the map *is* the territory, not of course in terms of the first condition on representation (that the relationship of 'standing for' is recognizable), but rather in terms of the second condition on representation, since what is *intended* to be represented is recognizable only by reference to the procedures for recognizing what is *actually* represented.[13]

This is not the case for discursive representations. As Putnam's examples shows, it is possible for a listener successfully to identify what the speaker is intending to represent, even while having few or no procedures by which he or she would be able confidently to recognize what is actually being represented. This can equally be the case for 'concrete' names, such as 'obsidian'; 'abstract' names, such as 'diffarreation'; and names for 'imaginary' entities, such as 'complex number'.[14] There are, to be sure, limiting cases (the reader may just have encountered one or more) in which referential or representational intent fails utterly, inasmuch as the interpreter has absolutely no idea what the producer is 'talking about'. Nonetheless, the general point remains: the procedures underlying the recognition, in discourse, of that which is intended to be represented, are relatively autonomous from the procedures for the recognition, outside discourse, of that which actually is represented.

It is important to emphasise, here, that by characterizing the relationship of one set of procedures to another as 'relatively autonomous', I do not intend to mean that they are wholly independent. In fact, what underlies and permits this partial (and permeable) dissociation of recognitory procedures for 'things', and recognitory procedures for 'discursive concepts', is the (equally partial and permeable) distinction between 'linguistic knowledge' and the 'assumptions', to use Sperber and Wilson's terminology, that provide a *background* to the understanding of utterances in language. This notion of background is discussed in the following long quotation from Searle (1986):

> Take the sentence: George Bush intends to run for the presidency. In order to fully understand that sentence, and consequently, in order to understand a speech act performed in the utterance of that sentence, it just isn't enough that you should have a lot of semantic contents that you glue together. Even if you make them into big semantic contents, it isn't going to be enough. What you have to know in order to understand that sentence are such things as that the United States is a republic, they have an election every 4 years, in these elections there are candidates of the two major parties, the person who gets the majority of the electoral votes becomes president. That list is indefinite, and you can't even say that all the members of the list are absolutely essential to understanding the sentence; because, for example, you could understand that sentence perfectly well even if you didn't understand about the electoral college. But there is no way to put all of this information into the meaning of the word *president*. The word *president* means the same in "George Bush wants to run for president" and in "Mitterand is the president of France". It isn't that there is some lexical ambiguity over the word *president*, it is just that the kind of knowledge you have to have to understand those two utterances doesn't coincide. All of that network of knowledge or belief or opinion or presupposition I want to give a name to; I call it the "network". If you try to follow out the threads of the network, if you think of what you would have to know in order to understand the sentence "George Bush wants to run for president", you eventually reach a whole lot of stuff that looks really weird if you try to say that they [*sic*] are part of your knowledge or belief. For example, you will get to things like: people generally vote when conscious, or there are human beings, or elections are generally held at or near the surface of

the earth. These propositions are not like the genuine belief I have to the effect that larger states get more electoral votes than smaller states. In the way that I have a genuine belief about the number of electoral votes that goes to the state of Michigan, I don't in that way have a belief that elections go on at or near the surface of the earth. If I was writing a book about American electoral practices, I just wouldn't put that proposition in. Why not? Well in a way, it is too fundamental to count as a belief. It functions rather as part of the background stance that I take towards the world. There are sets of skills, ways of dealing with things, ways of behaving, cultural practices. The fact that part of my background is that elections are held at or near the surface of the earth manifests itself in the fact that I walk to the nearest polling place and don't try and get aboard a rocket ship or something like that. The fact that the table in front of me is a solid object is not manifested in any belief as such, but rather in the fact that I'm willing to put things on it, or that I pound it, or I rest my books on it, or I lean on it. Those, I want to say, are stances, practices, ways of behaving. This then for our present purposes is the thesis of the background: all semantic interpretation, and indeed all intentionality, rests on a background that does not consist in a set of propositional contents, but rather, consists in presuppositions that are, so to speak, preintentional or prepositional (pp. 16–17).

Searle's argument is clearly rooted within a variant of epistemological pragmatism. I shall take up some of the epistemological issues in a moment, and for now let us merely note some of the issues which his analysis throws up, and which appear to be as problematic as the 'standard semantic' position of which he offers a critique. What, for example is the nature of 'background', if it is not knowledge or belief? What is a 'prepropositional presupposition'?

Before addressing these questions, I wish to attempt to reformulate, in the light of the foregoing, the classical linguistic notions of 'sense', 'reference', 'meaning', 'denotation' and so forth, in such a way that they tie in with my general analysis of representation.[15]

Sense The sense of a term is its content as a discursive concept, and is that which enables it to fulfil, in discourse, the conditions on representation. The sense of a term (or expression) is thus its representational meaning, as defined above.

Semantic value The semantic value, or lexical meaning, of a term is that aspect of its sense which relates it to other senses (discursive concepts) within a language, and governs the lexico-grammatical distribution of the term (in the classical formulation, in terms of the paradigmatic and syntagmatic relations which the term contracts with other terms).

Denotation The denotation of a term is that aspect of its sense which relates it to the recognitory procedures associated with what is definitionally, canonically or prototypically represented by the term. In the theoretical position which I am advocating, denotation is a language-to-world relation, independently of whether the term is, in traditional parlance, analytic or non-analytic.

Reference The reference of a term, for an interpreter on a given occasion of discourse, is that which is identified by the term by virtue of the fulfilment of the conditions on representation.

Signification The signification of a term, for an interpreter on a given occasion of discourse, is the representational meaning of the term plus its contextual meaning.

Representation, re/cognition, relativity

The above definitions emphasise the essentially socio-communicative nature of representational meaning in all its aspects. A discursive concept, on this account, and in line with Frege's theory, is something which is 'there' in the language system, to be grasped, entertained, and utilised in discourse; it is, emphatically, *not* a psychological 'copy' of the signified—sense is not a matter of similitude. On the other hand, this account departs from Frege in resolving sense and reference in terms of communicative and representational *adequacy*, and not in terms of truth. 'Truth' does figure in the account, inasmuch as denotation implicates the linkage of 'representation' with 'reality', or at least with our procedures for recognizing (or misrecognizing) reality. However, as I shall argue, such linkages are grounded not in a pre-ordained, or empirically 'similized', correspondence between reality and representation, but in organized social practices, practices furthermore which are oriented to the active transformation of, and adaptation to, the world in which we exist as knowing subjects.

Let us consider, for example, a kind of 'misrecognition' which differs from the misrecognitions discussed previously, in that the conditions on representation are preserved, rather than violated. Such is the case for any material, social or natural form which remains in some way opaque to the subject; yet whose opacity cannot be attributed to the intentions of any individual or group designer, but exists by virtue of the exteriority of the form to all subjects, and its 'domination' of their practices. Such for example is Marx's analysis of the Commodity form in volume 1 of *Capital*. Marx writes: 'Whence, then, arises the enigmatic character of the product of labour, as soon as it assumes the form of a commodity? Clearly, it arises from this form itself' (Marx, 1976 [1867]: 164). Marx then intriguingly characterizes the nature of this form as 'socio-natural', having a structure which induces in the mind of the subject a deceptive yet motivated misrepresentation of the object: 'It is nothing but the definite social relation between men which assumes here, for them, the fantastic form of a relation between things' (p. 164).

Except in the case, which I shall exclude, of conspiracies, in which the representation of an 'interest' is deliberately and mendaciously concealed,[16] ideologies, by which I mean 'false' or 'not wholly true' representations in general, are 'misrepresentations' which arise not from the intentions of subjects, but by virtue of the structure of that which they represent, and the *relationship* of that which is represented to the representative subject.

Because 'the world' does not spontaneously reveal its 'true' structure, subjects represent it in terms which satisfy empirio-practical commonsense, or the constraints of our everyday 'form of life' (Wittgenstein, 1953), rather than in terms which satisfy material determinations at a 'deeper' level. Thus, to use Searle's example, we commonly represent, to ourselves and to each other, tables as being solid bodies, even though many of us know that this representation is, at a more 'basic' level, misleading. Ideologies arise, then, spontaneously from our joint and individual interactions with the world, rather than from any prior intentional, representational act by another party. Unlike in cases where the conditions on representation are not attained, the 'intersubjective flow' of ideological representations is not interrupted, but sustained, or at least sustainable; but that which subjects take the thing to be is not, or is not wholly, what in fact it is—appearance and reality are out of joint, but this disjointedness is itself concealed, by virtue not of intention, but of reality/representation.

Ideologies can be virtuous as well as vicious. In particular, the kind of empirio-practical ideologies which govern our everyday life and communication—such as the 'belief' that the world consists of discrete solid objects—are arguably *necessary* modes of representation. This idea, of course, goes back at least to Kant—and contemporary theories of 'ecological optics' (Gibson, 1979; see also Chapter 4) may be read as an argument that organisms such as humans are biologically adapted to process precisely that information that is afforded by the world under the aspect of its segmentation into discrete solid objects. Ideologies, in general, tend to be not so much 'false theories' as partial, or one-sided, representations, their 'one-sidedness' consisting in the accentuation of some aspects of 'reality' and the suppression of others.

This is arguably the case even for the most patently vicious ideologies. For example, the word 'Jew', uttered by a Nazi or other anti-Semite, does serve to sustain shared reference to a certain group of individuals; and the canonical criteria for the denotation of the term will, at least in part, overlap with those employed by other speakers who do not subscribe to anti-Semitic ideologies. What other speakers would usually wish to say about references to Jews by a Nazi is that is embodies a *misrepresentation* of the Jewish people in a quite different way than that involved in either failures of recognition, or failures to understand the representational status of a term. We would say that a Nazi has a bizarre, pathological and false 'idea' of Jews, embedded within a vicious and false system (background) of beliefs; or that the *signification* of the word 'Jew' for a Nazi is pathological and out of line, both with 'reality' and with other speakers' understanding of it.

This example can help to clarify both the notion of 'one-sidedness', and the notion of 'prepropositional presuppositions'. Most of us would agree that the *presuppositions* involved in the use of the term 'Jew' in a speech by Hitler are not 'one-sided' in the sense of being 'partly true'. We would rather wish to say that they were *false*, if they were actually stated as propositions. Thus, the 'misrepresentations' involved in Nazi ideology consist not in the 'half-truth' of what is presupposed by the functioning of the term 'Jew' in its discourse, but precisely in the representation of the Jewish people under the aspect of those presuppositions which, if stated as propositions, would be false.

In more benign ideologies, we would perhaps wish to say that, depending on *how* the presuppositions were actually propositionalized (recall the arguments above regarding the different ways of representing through propositions), the resulting propositions might be either true or false—viz. 'Tables are solid objects within whose spatial boundaries there is no empty space' vs. 'Tables are solid objects, and not gaseous or liquid'—or, even more loosely, cannot be shown to be true, viz. 'This table is the same object as it was an hour ago'.

It seems then that the nature of ideologies, vicious and virtuous, consists in the representation of reality under the aspect of certain presuppositions at least some of which, when propositionalized, are either false or cannot be shown to be true; and that there is a motivated relationship between these presuppositions and the 'meaning' of the segment of reality under representation. In this respect, the relationship between the representing subject and the represented object, in ideologies, is constituted precisely by those presuppositions which are generated 'spontaneously' in the subject by virtue of the *significations* of the object (including its place in other discursive representations and signifying systems). It is in this sense that it can be said that ideologies 'dominate' or 'constitute' subjects (Althusser, 1971), insofar as the objects which fall under ideological representations, as well as the discursive and non-discursive practices within which such objects are embedded, pre-exist subjects, constraining subjects' interactions with, and interpretations of, them.

From this point of view, ideologies are not only 'necessary' (in the benign sense) for our everyday dealings with the world, but are also inevitable accompaniments of *any* (everyday or scientific) practice, on the assumption that we can never attain Godlike absolute knowledge. For we can never, in principle, be sure that any representation which we utilize in our commerce with the world does not fall under the aspect of some presupposition which, if stated as a proposition, would be false. Does this mean we should be sceptical of all claims to knowledge, or suppose that there is no good reason for believing one representation to be better than another?

This is, of course, the position of epistemological relativism, and I wish briefly to advance some arguments against it. To begin with, the fact that a representation is ideological insofar as it falls under the aspect of a presupposition which, when propositionalized, is false, does not mean that the representation is thereby itself wholly false, or that it can be dispensed with. Take, for example, the term 'atom'. The term is old, and our beliefs regarding its denotation have changed radically. One of the most important developments in physics was the realisation that atoms possess internal structure, and are decomposable, both in principle and in practice. Consequently, the sense of the term 'atom' had to change to accommodate the new information that what was previously presupposed by its usage, that it represented the ultimate and indivisible unit of matter, was false. This change did not present insuperable difficulties.

The point is simple: representations can change, and in particular what is presupposed by a representation can change, with a consequent gain in the adequacy, utility and 'veridicality' of the representation, irrespective of the fact that the initial conditions for the production and circulation of the representation may have included false presuppositions.

However, to say that a representation has changed, is not to say that it has changed *beyond recognition*. The sense of the term 'atom' has continued, through time, to enable us to recognize that it stands for the object which it does stand for. Further, this is dependent upon the term having a denotation—upon 'us' (someone, somewhere, as Putnam insists, in the linguistic community) having procedures for recognizing atoms. Certainly the procedures which must be employed to recognize atoms are more complex than those involved in recognizing, say, tables—they amount to complex and highly structured social practices. Such practices are problematic, insofar as they too depend upon a large set of presuppositions about the nature of reality. In this sense, our experimental and cognitive 'grasp' upon atoms is rather more vulnerable to disconfirmation than our 'grasp' on everyday phenomenal forms.

On the other hand, the very range and structural complexity of the presupposition-set involved in 'our' theoretical and practical dealings with atoms, serves to secure the representation of atoms by 'atoms' against a too-easy disconfirmation. The sense of 'atom' can, within limits, be accommodated to incorporate changes in its associated presupposition-set, provided such changes do not render, all at once, all the procedures for recognizing atoms invalid.[17] The process of accommodation of the sense of 'atom' goes in parallel, as it were, with the process of accommodation of organized social practices whose object is to manipulate atoms and increase 'our' knowledge about them.

Thus, the term 'atom' possesses *sense continuity* across changes in its associated presupposition set. This is a strong argument against treating such presuppositions as 'criterial', or as 'entailments' in the analytic sense. Further, the sense continuity of 'atom' is closely associated with some degree of continuity in the recognitory procedures applied to atoms—there is a complex dialectical interplay between knowledge, practice and representation.

Sense continuity serves to secure representations—and the discourses within which they are embedded—against disruption by new information to be added to denotations. The way in which this 'discursive buffering' operates can metaphorically be described as involving the insertion of contextual 'as if' clauses, between background presuppositions and foreground propositions. Such 'as if' clauses serve not to foreground the presuppositions themselves, but rather to acknowledge both the possibility that such a foregrounding *might* be operated, and the resultant propositions tested; and the inevitability that in at least some cases the product of such a test will be the conclusion that the proposition is false. So, for example, 'we' view Newtonian space as a special case, holding within strict contextual limits, of the properties of space as conceived in the theory of relativity. Thus, the propositionalized part of a representation is *local.* The local nature of propositional truth is explicit only to the extent that we know, and can propositionalize, the 'as-if' clauses intervening between proposition and presupposition. There is also, however, an *implicit* sense in which propositional truth is local, insofar as further knowledge may require the insertion of further (as yet unknown) 'as-if' clauses.

The necessarily local nature of propositional truth is not, however, a *merely* relative truth, for its necessary degree of relativity is, implicitly or explicitly, *recognized* and *represented* in the form of 'as if' clauses which, far from being arbitrary, are simply the form assumed by the limits of actual knowledge and practice—and hence are (from an epistemological point of view) a part of the general conditions on representation.

The epistemological position outlined above is not relativistic, inasmuch as the socio-communicative conditions on representation are paralleled by the socio-practical conditions on the adequacy of the denotational components of discourse. Because discursive concepts are understood to be *objects* of cognitive and communicative processes,[18] the theory I have proposed can be considered to be *realist* in the Fregean sense; that is, in opposition to 'psychologistic' theories according to which discursive concepts are subjective, mental entities.

This epistemological position is not to be understood, however, as a 'classical realist' one, if by that is understood a 'correspondence' theory of truth; such correspondence as exists between 'representation' and 'reality' is given not by verisimilitude, but by the (socio-pragmatic) conditions on the production and circulation of representations. Since the denotational aspects of discourse are themselves subject to the general conditions on representation, as well as to non-discursive constraints on the adequacy of social practice, it follows that 'truth', on this account, while it remains a relation between discourse and what is external to discourse, is nonetheless fundamentally discourse-dependent in its propositional formulation.

In this respect, the analysis I have offered implies that the only access we (as a linguistic community) have to reality is through language and discourse, which provides the mediating and signifying structures permitting the production and circulation of representations. This statement needs, however, to be qualified in the light of considerations regarding 'background assumptions'.

Such background assumptions include such information as underpins the relationship of cognitions to the phenomenal world: in the terms of classical philosophy, those 'intuitions' which enable us, for example, to recognize a table as an appropriate supporting surface on which to put keys, without entertaining the proposition 'That is a table'. Like any other species, human beings are adapted to an environment within which their behaviour is largely governed by 'simple' cognitions and recognitions, which do not necessitate continuous recourse to discursive representations or propositional attitudes.[19] Furthermore, the force of the analysis of 'background' offered by Searle is that it is, in principle, impossible to draw a sharp dividing line between discursive representations (such as might be necessitated for the comprehension of the utterance 'you'll find your keys on the table in the hall'), and non-discursive recognitory procedures, or, more generally, 'stances' for behaving in the world.

If, as I have suggested, it is implausible rigidly to separate representational meaning from 'meanings' in the world around us—that is, the signifying structure of the world as it presents itself to us—then we must conclude *either* that representational meaning reduplicates, in some kind of internal 'language of thought', those 'meanings in the world'; *or* that some kind of continuity exists between representational meaning and 'meaning in the world', such that representation itself achieves some material instantiation *in the world*. Searle, in the quotation earlier in this section, advances some arguments against the former hypothesis, and much of the rest of this book is devoted to arguments for the latter hypothesis, in the form of the theory of the materiality of representation.

Without anticipating these arguments, what I wish to emphasize is that the general epistemological position argued for here is not only realist, but also naturalistic, in the sense that it advances an account of representation which stresses the continuity and inter-relatedness of adaptation, transformation, cognition and representation. Because of the fundamentally social matrix within which this developmental and constructivist process takes place, this is at least one way in which the concept of a 'socio-naturalistic' approach may be understood.

Psychological concepts

It was stressed above that representations, understood as discursive concepts, should not be equated with mental copies, or images. Indeed, discursive concepts should not be thought of as 'mental' entities at all, in the usual sense in which the term 'mental' equates to 'subjective'. Obviously, because of their mediational role in communication and representation, discursive concepts possess a cognitive or mental aspect; equally, though, they posses a material aspect inasmuch as they are embodied in signifying structures, ranging from 'raw' objects to linguistic signs. The cognitive aspect of discursive concepts should, according to the theory I have proposed, be thought of as *inter*-subjective, or inter-mental; and as being underpinned by material social practice.

Nonetheless, not only can there be no representation without signification, but there can quite evidently be no representation without a *subject* of representational intention and intepretation. The question thus arises of the nature of subjects' *individual* mental representations of discursive concepts—that is, of *psychological concepts*, in the sense in which we can speak of 'my' or 'your' concept of alabaster, bimetallism, the second law of thermodynamics, or whatever.

Only rarely in the history of thinking about concepts have attempts been made[20] to discuss both the 'discursive' and 'psychological' faces of conceptualization simultaneously. Mindful of the snares of 'psychologism' and 'conceptualism', philosophers have tended to neglect psychological issues entirely, or to be content merely to distinguish discursive 'meaning' from private 'idea'. Psychologists, notoriously, have tended to play fast and loose with philosophical distinctions, and to fall back, in the face of difficult questions, on naïve and largely inexplicit empiricist assumptions about 'abstraction' and 'classification'.

In this section, I do not wish to do more than lay down a few signposts, indicating the direction of subsequent discussion. The first observation along this path is that arguments which are valid in respect to one face of the Janus-like notion of 'concept' may not hold for the other face. An important example here is that relating to imagery. While I have emphasized that discursive concepts should not be seen as 'copies or 'images', such philosophical arguments against a similitude theory of reference quite rightly cut no ice with cognitive psychologists who are aware of the significant and vital role of imagery and 'mental models' in thinking (Johnson-Laird, 1983; John-Steiner, 1987).

I have emphasized the 'permeable' nature of the boundary between 'linguistic knowledge' and 'background assumption', such that the deployment of discursive concepts in actual discourse implicates (psychological) procedures for recognizing 'things', while not being reducible to the latter. By the same token, it is important to recognize the *polyvalence* of psychological concepts, in which the psychological acquisition of aspects of 'sense' (i.e. denotation, semantic value) depends upon a rich and seamless web of linguistic and non-linguistic social practices. Developmental psychologists have, in recent years, emphasized the importance of interpersonal contexts of 'joint action' in cognitive and linguistic development, rendered in such terms as 'scaffolding' (Bruner, 1983), in which the developing child is engaged in a process of 'guided re-invention' (Lock, 1980) of language. The acquisition of psychological concepts, on this account, is a matter of the personal *appropriation* (Leontiev, 1981) of discursive concepts and of the conditions on the possibility of their employment. Such a view does not specify in advance the phenomenological structures of mental representation, and still less does it preclude imagery playing a role in individual mental processes.

This account does, however, stress that the 'having' or 'entertaining' of a concept is not an all-or-nothing affair, but a *developmental process*. In other words, I shall follow the assumption of the classical genetic psychologists, such as Piaget and Vygotsky, that concept formation, or concept learning, occurs; and that there exist certain cognitive prerequisites for the psychological appropriation of discursive concepts.

The most obvious prerequisite for the acquisition of concepts is a grasp of the conditions on representation, as I have defined them. In previous sections I have analysed cases of failure to meet the conditions on representation—including cases, such as *trompe l'oeil*, in which a representation is mistaken for a real object. Whereas, with *trompe l'oeil*, this mistake is attributable to the intentions of the producer, psychologists have often been concerned with cases in which a similar mistake appears to be attributable to a lack of cognitive capacities on the part of the interpreting subject.

As an example, think of a kitten darting behind a television set to 'find' a moving object on the screen. In most, if not all, non-human species, the disappearance with age of this kind of 'mistake' in behaviour can be attributed, not to the maturation of a representational capacity, but to a simple process of dishabituation: eventually, the kitten will come to perceive the pictures on the television screen (if it bothers to look at them at all) as mere patterns (motifs), and to discriminate them from other objects. In the case of human infants, although we can sometimes observe similar behaviours to that of the kitten (for example, an infant may attempt to grasp a brightly coloured image in a picture book), the disappearance of these is usually accompanied by an evident sophistication in understanding representation. An eighteen-month-old child, we may suppose, when pointing to a picture of a train in a book and uttering the expression 'train', is demonstrating her mastery of the conditions on representation, not mistaking the 'represented' train for a 'real' train.[21]

If, however, it is easy to find cases where it seems safe to attribute to children an adequate working knowledge of the conditions on representation, it is equally easy to find examples which seem (to some observers) to indicate the opposite. Consider, for example, the following observations from Piaget (1977b [1927]) and Vygotsky (1986 [1934]):

> During the earliest stages, the child believes that he thinks with his mouth, that thought consists in articulating words, and that these words themselves form part of the external things. The voice, being thus identified with thought itself, is regarded as a breath which participates with the surrounding air, and some children go so far as to say that it is identical with the wind in the trees, and that dreams are made of "wind". They are quite incapable of distinguishing between thought and the things thought about. To use the expression chosen by M. H. Delacroix, the sign "adheres" to the thing signified ... Word and name are about all that the child knows of thought, since he identifies thought with the voice. Now, names are, to begin with, situated in objects. They form part of things in the same way as does colour or form. Things have always had their names. It has always been sufficient to look at things in order to know their names. In some cases, this realism turns to magic: to deform the name is to deform the thing (Piaget, 1977b: 131).
>
> The word, to the child, is an integral part of the object it denotes. Such a conception seems to be characteristic of primitive linguistic consciousness ... Simple experiments show that preschool children "explain" the names of objects by their attributes. According to them, an animal is called "cow" because it has horns, "calf" because its horns are still small, "dog" because it is small and has no horns; an object is called "car" because it is not an animal. When asked whether one could interchange the names of objects, for instance call a cow "ink", and ink "cow", children will answer no, because "ink is used for writing, and the cow gives milk" ... We can see how difficult it is for children to separate the name of an object from its attributes, which cling to the name when it is transferred like possessions following their owner (Vygotsky, 1986: 222–223).

Both Piaget and Vygotsky see, in this well-attested phenomenon of 'verbal realism', evidence that the child is beset by 'magical' and 'prelogical' thinking, a condition which for Piaget was a part of a global 'egocentrism', and which both writers believed the child shared with 'primitive' peoples and the deranged:

> [Autistic thought] is the thought of the child, of the neurotic, of the dreamer, of the artist, the mystic. It has also been studied by Lévy-Bruhl under the name of *prelogical thought*, of which the principal characteristic amongst primitive

> peoples is its fusion with magic (Piaget, 1977b [1920]: 56). It is therefore our belief that the day will come when child thought will be placed on the same level in relation to adult, normal and civilized thought, as "primitive mentality" as defined by Lévy-Bruhl, as autistic and symbolical thought as described by Freud (Piaget, 1977b [1924]: 117).
>
> There is another very interesting trait of primitive thought ... This trait—which Lévy-Bruhl was the first to note in primitive peoples, Alfred Storch in the insane, and Piaget in children—is usually called *participation* ... Since children of a certain age tend to think in pseudo-concepts, and words designate to them complexes of concrete objects, their thinking must result in ... bonds unacceptable to adult logic. ... Primitive peoples [also] think in complexes, and consequently the word in their languages does not function as the carrier of a concept but as a "family name" for groups of concrete objects belonging together, not logically, but factually ... Storch has shown that the same kind of thinking is characteristic of schizophrenics, who regress from conceptual thought to a more primitive level ... Schizophrenics ... abandon concepts for the more primitive form of thinking in images and symbols. The use of concrete images instead of abstract concepts is one of the most distinctive traits of primitive thought. Thus the child, primitive man, and the insane ... all manifest participation—a symptom of primitive complex thinking and of the function of words as family names (Vygotsky, 1986: 128–130).

There are two noteworthy aspects to this argument. The first is that it is couched in terms of a general theory of development and evolution: consisting in the equation of 'childish' with 'primitive', 'illogical', 'irrational' and, frequently enough in some versions of the theory, such as that of Freud, 'female'. In Chapter 3, I shall examine the role of this 'Phylocultural Complex' in theories of symbolic evolution in more detail. The second aspect of the argument, more pertinent here, is that it locates the nature of this illogicality in a defective comprehension of the function of 'representation', whereby the 'signifier' is wrongly believed to be an intrinsic property of the 'signified'; and this defective comprehension is itself seen as part of an overall tendency in 'primitive' thought towards a (misplaced) concreteness and a belief in the potency of the symbolic. In this explanation, we see not only a general theory of 'primitive mentation', in which signifieds are collapsed into signifiers, but also (in the theoretical apparatus itself) the familiar collapsing of representation into signification. In the work of Piaget, this takes the form of a 'special theory' (within the 'general theory', which is examined in more detail in Chapter 3) in which mental representation is seen as being grounded in a unitary *semiotic function*.

There is, however, an alternative explanation for the phenomenon of 'verbal realism'. That is, that in answering a difficult question such as 'why is a cow called a cow?', pre-school children resort, not to the far-from-spontaneous, contentious theory of the arbitrariness of the sign, but to a spontaneous and reasonably insightful account of how they might go about determining whether such-and-such an object may appropriately be referred to by a certain word. Thus, cows are cows, and may be referred to as such, by virtue of their having horns, and so on. Further, their account is informed by an implicit understanding of the contextual determination of reference by the set of available contrasting items—calves, cars and so on (see Freeman *et al.*, 1982).

Which of these explanations of 'verbal realism' is correct, and indeed whether either of them is correct, is of course in part an empirical matter (see also Markman, 1976). However, the example serves to highlight an important theoretical issue. This concerns the extent to which mental processes in young children are qualitatively different, as Piaget and others have believed, from those of adults. Recently, developmental psychologists and philosophers of language have advanced a number of important arguments against this position (see, for example, Donaldson, 1978; Fodor, 1976; Macnamara, 1982), or at least against the too-ready assumption of its being the case. In this book I shall be arguing for a developmental theory, but not one in which 'illogicality' or 'egocentrism' play central roles, either as symptoms or as explanations of the mentality of children.

Propositional attitudes

The introduction of psychological concepts into this discussion inevitably raises the vexed issue of the nature of propositional attitudes and mental representations. Propositional attitudes are, almost by definition—that is, unless one adopts the behaviourist approach which views them as dispositions to behave—mental states. In the current parlance of cognitive science, this is taken to mean that they are 'representational' in the specific sense of *individual mental* representation. Psychological concepts, as discussed in the preceding section, are mental representations in precisely the same sense (see Fodor, 1981: 259). From these premises—which for the moment I shall take as given—it has been argued (notably by J. A. Fodor) that the *objects* of propositional attitudes are themselves mental representations, consisting of propositions formulated in an internal 'language of thought', or 'mentalese'. As Fodor (1981: 187) puts it 'propositional attitudes are relations between organisms and formulae in an internal language; between organisms and internal sentences, as it were.'

Fodor summarizes his position, which he dubs 'The Representational Theory of Mind' (RTM), and explicitly situates within the tradition of empirio-rationalism, as follows:

> (a) Propositional attitude states are relational.
>
> (b) Among the relata are mental representations (often called "Ideas" in the older literature).
>
> (c) Mental representations are symbols: they have both formal and semantic properties.
>
> (d) Propositional attitudes inherit their semantic properties from those of the mental representations that function as their objects ... and (presumably) the semantic properties of the formulae of natural languages are inherited from those of the propositional attitudes they are used to express (Fodor, 1981: 26, 31).

The Fodorean argument from this position to the neo-rationalist conclusion that all concepts are innate is well-known, and I shall not rehearse it here (see Chapter 4). I shall take it, however, as given that *if* one accepts propositions (a) to (d) above, Fodor's argument to the innateness of concepts is compelling and, perhaps, inevitable. What I wish to argue, though, is that, fortunately for developmental psychology, it is not necessary to accept these propositions, and indeed that they represent an erroneous theory.[22]

The essential fallacy of RTM, I shall argue, is the assumption that the objects of intentional stances (including not only propositional attitudes, but the having or entertaining of concepts) are mental representations or ideas. It will immediately be objected that, having conceded that mental representations exist (as psychological concepts), and that propositional attitudes (and other intentional stances) are mental (representational) states, this leaves (a) no other apparent function for psychological concepts (b) no other apparent explanation of what the objects of intentional stances *are*.

Two possible ripostes to objections (a) and (b) may be dealt with summarily. First, I do not suggest that mental representations are mere epiphenomena or *Scheinbilden*, along behaviourist lines, any more than I suggest that propositional attitudes are dispositions to behave (verbally or otherwise). Second, I do not suggest that the objects of intentional stances are objects in the world or *representanda*; that is, I accept that they are *representamina*, or signs. In other words, if I were to say 'think of the desert', I could perfectly well be intending that you entertain the concept of 'desertness', and not intending that you direct your thoughts to a particular desert. Equally, for example, if I believe that Martians control my body, then the object of my belief is a *representation* in which the concepts 'Martian' and 'my body' figure—not Martians, either under representation or in the raw.

Since, so far, this seems only to compound the contradiction, it is as well at this point to articulate the alternative to RTM which I wish to advance. In essence, my proposal[23] is that the objects of intentional stances are *discursive*, and *not* psychological, concepts; and that it is the acquisition of natural language (not the innate knowledge of mentalese) that enables subjects to master (or appropriate) discursive concepts and adopt intentional stances.[24]

Nevertheless, the relationship between subjects and discursive concepts is not an immediate one. Indeed, one is tempted to ask how it possibly could be, without the re-postulation of the innateness hypothesis. Discursive concepts are grounded in discourse, and this is in turn grounded in social (and in private) activities and practices, into which discourse introduces new levels of signification, thereby rupturing, displacing and re-constructing the fabric of the social and the impersonal world alike. Such displacements and re-organizations are also constitutive of the subject and his or her psychological capacities—through and in discourse, subjective, psychological concepts are articulated with the significant structure of the world (see Chapter 5) as internal *cognitive signs* enabling the subject to adopt stances and positions within discourse, and in respect of the world. Psychological concepts, as Vygotsky always emphasized, are signs, but they are not copies, or surrogates, of discursive concepts, any more than discursive concepts are surrogate 'things'.

With the abandonment of the notion of the 'idea' (in RTM, 'representation') as isomorphic with both discourse and the world, we are led back to the notion (attributable to both Piaget and Vygotsky) of conceptualization, not as similitude and/or abstraction, but as *activity*, and the *punctuation* and *organization* of activity.[25] Psychological concepts, then, are psycho-semiotic structures which mediate the subject's discursive practices, and which are constructed in the process of socio-cognitive development. These processes, and the articulation of psychological concepts with the significant structure of the world, form the subject matter of Chapter 5. What, though, of propositional attitudes?

At this point, we may return to the larger issues of representation upon which this chapter has focussed. Representation (as all theories which admit intentionality recognize) is representation *of* something, and this holds equally for the discursive representations which I have postulated to be the objects of propositional attitudes. Furthermore, one of the conditions on representation is, as we have seen, that the object (or state of affairs) under representation should be recognizable to an interpreter.

What I shall suggest is that propositional attitudes may best be thought of as relations of *commitment* (including negations, epistemic modals, disclaimers, etc.) to states of affairs in the world. Such commitments, in order to have pragmatic force, must be discursively represented. On this account, a propositional attitude is not a relationship between an organism and a mental representation, but a relation between a subject and a (discursively) represented state of affairs, such that both: (i) the state of affairs, and (ii) the nature of the subject's commitment to that state of affairs are (in principle, and given that the propositional attitude receives linguistic expression) recognizable to an interpreter.[26] Propositional attitudes, on this account, are communicatively and pragmatically engendered; they are not so much 'things' you 'have', as acts which you (in principle) commit. To entertain a propositional attitude is to entertain the possibility of acting socially, by making a commitment to which you may, in the terms of Shotter (1984), be held accountable; and to express a propositional attitude is to make a speech-act which is both performative, in making that commitment public; and appellative, in that it 'calls to mind' the commitments of the speaker, and the states of affairs to which they are directed, for the benefit of the hearer.

Explicated propositional attitudes are thus socially constitutive discursive practices, constituting the subject in a particular relation to the world and to others. There is, however, no principled dividing line between the *content* represented in explicated propositional attitudes, and that implicit in pre-propositional, presupposed background: the difference is, on the one hand, precisely one of explicitness, and on the other hand a discursive-practical difference in terms of commitment and accountability. The 'explication' of foreground from background is thus a developmental process both in the time-scale of discourse, and in the time-scale of ontogenesis.

Mental representations—cognitive signs—are necessary, to be sure, to the entertaining of propositional attitudes: but they are secondary to, not the origin of, the discursive concepts actuated in natural language, which the subject must appropriate in order to perform communicative acts. Cognitive development, in this theory, is the process of appropriation of discursive concepts and positionings—and is a process open to empirical study in precisely the way which Fodor's theory denies. Furthermore, the theory which I have outlined leads naturally to the hypothesis that the process of appropriation of discursive concepts is articulated upon the *material representations* of interleaved discursive and non-discursive practices, as realized in concrete discourse and in artefactual objects.

Notes

1. From the Greek *kanon* = rule.
2. Thus, syntactic rules in the Chomskyan sense are not instances of canonical rules, since violations of them in performance are in some sense 'random' perturbations; should they be 'systematically distorted' in their instantiation, then we should have to say that the rules (which *are* constitutive, for Chomsky, of competence) had *changed*. This is not the case for canonical rules.
3. See also Freeman, 1980 for an extended discussion of the notion of canonicality.
4. *The Rider* by B. A. van der Leck.
5. *Trompe l'oeil* was not only known to the Ancients, but was used as a pedagogic example in classical philosophies of art (Bernheimer, 1961: 7ff.).
6. Cited in Lyons, 1977: 53.

7. This was of course the starting position adopted by Wittgenstein in the *Tractatus*, and which he subsequently criticised as naïve. It is not clear, however, that the latter Wittgenstein wished wholly to abandon the theory of representation set out in the *Tractatus*, or merely to renounce those aspects of it which depended upon the strict determinacy of propositional meaning. See Wittgenstein, 1961 [1921]; Fogelin, 1976.
8. See below for further discussion of the representation of requests.
9. In other respects, however, and particularly in terms of their suggestions regarding pragmatic and logical inference, their approach is at variance with the one taken here.
10. Sperber and Wilson, 1986: 30. No sexist implications are intended by this example; to use the terminology of this chapter, if sexist inferences are drawn, they are drawn from and through complex signifying structures, not from what is represented.
11. See Sperber and Wilson, 1986, ch. 1, for an illuminating discussion of the vexed question of 'mutual knowledge'.
12. Correspondence is used here in the sense of isomorphism, rather than identity.
13. I am ignoring for convenience the fact that actual maps carry a great deal of coventionally symbolic information.
14. Thanks to John Shotter for the last example.
15. The distinctions drawn by Lyons (1977) and by Nelson (1985) have much in common with what I propose, but are not by any means identical.
16. 'Semiotics is in principle the discipline studying everything which can be used in order to lie' (Eco, 1976: 7).
17. This can of course happen, and did, for example, in the case of 'phlogiston'. Even here, notice that 'phlogiston' retains a *sense*, which includes the information that its denotation is non-existent and that the term is historical.
18. Strictly speaking, mediative objects, articulated through and in the mediation of both discursive and non-discursive practices with the fabric of the world.
19. A propositional attitude may be taken, pretheoretically, to be whatever is denoted by such expressions as 'the S that *p*, where S is a mental state term such as 'belief', 'desire', 'supposition' etc., and *p* is a proposition.
20. Peirce's work, from the standpoint of philosophy; and the theories of Johnson-Laird (1983), Macnamara (1982) and Nelson (1985), within psychology, are important exceptions; as is Fodor's work which is discussed below.
21. It might be objected that the mastery of these canons is incomplete. Concern is frequently voiced, for example, that children may take the 'represented' for the 'real' when viewing television, and believe that they can fly like Superman, or that the characters in a series 'really exist'. These are second-order issues, however, inasmuch as they relate to the distinction between 'factual' and 'fictional' representations and their narration; or to the tendency of children to emulate what they see represented despite their everday knowledge of what is possible or acceptable. In any event, these considerations apply to adults as well as young children, and while this does not mean they are unrelated to cognitive development, they are nonetheless

outside the scope of this analysis.

22. It should be noted that the innateness of concepts is not the only problematic consequence of Fodor's theory. Another is the logical regression involved in the notion of an organism having a relation to a mental representation—a notion which further necessitates some mental structure permitting the organism to have intentional attitudes independently of the content of the representations towards which they are directed.
23. The proposal is not, admittedly, entirely original, since it shares many of the assumptions of Fregean realism, amongst other positions. However, as discussed earlier, the socio-pragmatic cast of the theory distinguishes it from Frege's account.
24. This implies of course that the objects of propositional attitudes are, in principle, expressible as sentences in natural language. I assume with other accounts that the belief that 'it is raining' is more or less equivalent to the belief that 'il pleut', and that the more-or-less-ness of this equivalence is determined by sense or representational meaning, as I have defined it on pp. 57–58. As a further twist to the Quinean indeterminacy problem, however, I shall argue that there is no reason to suppose that the ability to entertain 'fully-fledged' propositional attitudes is any more saltatory or absolute than the ability to entertain discursive concepts.
25. It would, perhaps, be better not to speak of *psychological* concepts as representations at all, but rather as cognitive signs (recall the earlier distinction between representation and signification) enabling the appropriation of discursive concepts (representations) by means of which the conditions on representation (as a socio-pragmatic activity) may be fulfilled. The notion of 'mental representation', however, is so deeply embedded in our current cognitivist discourse that such terminological quibbling would probably obscure more than it illuminated.
26. Recognizable, that is, in the special, partial-permeable sense analysed above; thus, the theory can deal adequately with the belief that one's body is controlled by Martians.

3 Evolution and Development: The Phylocultural Complex

Auguste Comte formulated stage-successive laws of history between bouts of clinical madness, and Manuel reports that "as his illness became aggravated, he had felt himself regress through various stages of metaphysics, monotheism and polytheism, to fetishism, and then, in the process of recuperation, had watched himself mount again through the progressive changes of human consciousness, at once historical and individual, to positivism and health."

Peter C. Reynolds (1981: 11)

Darwin and the mind of the child

The single most important factor in establishing child psychology as a field of scientific study was Darwin's theory of evolution. Indeed, probably the earliest 'modern' text of child psychology was Darwin's study 'A biographical sketch of an infant' (1877), which covered such areas as motor, emotional, communicative, intellectual and moral development. Darwin's most powerful and enduring influence on the emerging discipline of child psychology, however, was indirect, through such disciples as his friend and colleague George Romanes, to whom Darwin entrusted his unpublished psychological manuscripts, James Mark Baldwin, James Sully, Karl Groos and G. Stanley Hall. Whatever their other differences, all these psychologists shared a common commitment to, first, a comparative, historical, evolutionary approach to human psychology; and, second, a background in a scientific endeavour deeply influenced by the notion of 'recapitulation' as formulated by Ernst Haeckel (1874).

According to Haeckel[1]—one of the most energetic Darwinian propagandists in the late nineteenth century, as well as an influential evolutionary theorist in his own right—the developmental history of the individual organism, from embryo to maturity, repeats or recapitulates, in a reduced time-span, the entire evolutionary history of the species or 'race'. The human embryo, foetus and infant, for example, was said by Haeckel to pass successively through fish-like, amphibian, reptilian and mammalian stages, recapitulating the evolutionary stages preceding the emergence of the human species. Haeckel considered his theory of recapitulation to represent a 'fundamental biogenetic law'; and it was eagerly seized upon as a guiding principle for investigating the biological foundations of the human mind.

Ontogeny was seen as the key to unlock the secrets of 'mental evolution', including the evolution of language and symbolization. By understanding the mind of the child, psychologists hoped to gain insight into the 'childhood of the race', and to formulate general laws of mental and cultural development. Haeckel's 'biogenetic law' was not, however, the only, or even the foundational element in the 'Phylocultural Complex'—the set of assumptions dominating evolutionary and developmental studies at the turn of the century, based upon concepts of parallelism and repetition at different 'levels' of biological and cultural organization. Within the framework of the Phylocultural Complex, the mentality of the child was explicitly compared with the mentality of non-European peoples: the 'discovery' of the 'childish' mind, as an object of scientific inquiry, was more or less simultaneous with the 'discovery' of the 'savage' mind, and both (as we have seen in the previous chapter) were viewed as manifestations of a more general category of 'primitive mentality'.

Such 'child-primitive' comparisons (see Gould, 1977; Gardner, 1987: ch. 8) were clearly related to ideological justifications of ascendant European imperialisms, and to the role played by both anthropology and psychology in rationalizing the administrative measures requred for the regulation of 'backward' subject peoples.[2] They were also, however, both symptomatic of, and contributory to, the confusions and resistances incurred by the fundamental restructuring of European notions of time and history in the wake of the Darwinian revolution. In this chapter, I attempt to explore the role of the Darwinian and evolutionary legacy in general, and the Phylocultural Complex in particular, in the work of Freud, Piaget and (especially) Vygotsky.

Freud: Ursprache and the unconsciousness

Although Sigmund Freud (1856–1939) is best known as the founder of psychoanalysis, his concern with symbolic processes both predated his mature theory of psychosexual development, and encompassed a general preoccupation with the relations between biology, mind and culture. As Frank Sulloway makes clear in his biography *Freud: biologist of the mind* (1979), 'many, if not most, of Freud's fundamental conceptions were biological by *inspiration* as well as by implication ... Freud stands squarely within an intellectual heritage where he is, at once, a principal scientific heir of Charles Darwin and other evolutionary thinkers in the nineteenth century and a major forerunner of the ethologists and sociobiologists of the twentieth century' (p. 5). Sulloway argues that the 'Freud' of the Freudian myth is a 'crypto-biologist', inasmuch as the myth—of which Freud himself was the principal author as well as player—has sought to downplay or disguise the true extent of his intellectual debt to the psychobiological milieu within which his theories took shape.

The expression 'crypto-biologist' is a fortunate one, also, because, intertwined with the biological themes which inform Freud's work, are themes which can, from our historical vantage point, be recognised as semiotic: Freud was indeed a cryptologist, a decipherer and interpreter of the processes by which psychic symptoms arise as representations (representatives) of desires and memories censored from consciousness.

Any brief presentation of Freudian (or for that matter Piagetian or Vygotskian) developmental theory is necessarily selective as well as incomplete. Here I shall, for convenience, treat the three themes that I have identified above in turn: that is, Freud as cognitive theorist, Freud as evolutionist and Freud as semiotician.

In Freud's early (pre-psychoanalytic) work a persistent theme was already the attempt to understand the relationship between motivational-affective and perceptual-cognitive aspects of human psychology. This attempt, while clearly anticipated in Freud's work in the fields of neurophysiology and aphasiology, found its clearest expression in his (posthumously published) *Project for a Scientific Psychology* (1895). In this work, Freud attempted to conceptualize the emerging heuristic structure of psychoanalysis within a framework supplied by neurological theory. This line of attack, according to Sulloway, was subsequently abandoned as unattainable, and replaced by an evolutionary-cultural paradigm in Freud's later work.[3]

Nevertheless, as many authors have noted, (see for example

Wolff, 1960; Greenspan, 1979; Ingleby, 1983) a definite and dualistic theory of the relationship between affect and cognition underlies the entire Freudian approach to the ontogenesis (and phylogenesis) of symbolization, in which the tension between primary-process 'irrealism' or phantasy in the wish for gratification, and the reality principle resulting from the psychic structures engendered by secondary processes, remains perpetually and inevitably unresolved, while nonetheless providing a 'motor' for psychological development. In this respect, Freudian theory contrasts with Piaget's conception of symbolic and representational development, in which both cognition and affect are governed by the same processes of structural growth. The affect-cognition dualism, already present in Freud's early neuropsychology, was subsequently mapped by Freud (as might be expected, given traditional approaches to 'reason and the passions') onto the nature–culture distinction, and recast in the mould of evolutionary theory.

As Sulloway has convincingly demonstrated, evolutionary biological theory was centrally implicated in the later Freudian (psychoanalytic) project, in the form of the 'biogenetic-Lamarckian' synthesis. Neither Haeckel's 'biogenetic law' nor Lamarckian 'inheritance of acquired characteristics' are any longer seriously entertained in evolutionary biology,[4] the latter in particular having been proscribed by the twentieth-century neo-Darwinian 'synthetic theory' (see Chapter 4). Yet, during the period of both Freud's and Piaget's intellectual formation, these theories were not merely current, but orthodox: Darwin himself, for example, who did not live to see the rediscovery of Mendel's genetic experiments, accepted Lamarckian inheritance as a supplement to inheritance by natural and sexual selection.

The fusion of Lamarckian and biogenetic (recapitulationist) assumptions provided Freud with the rationale for his attempt, which became increasingly explicit in his writings, to adduce the phylogenetic history of culture through the study of the neuroses, and other manifestations of unconscious processes, such as jokes, slips of the tongue and, above all, dreams. Thus, Freudian 'cryptobiology' was simultaneously Freudian phyloculturalism. References to phylogenesis recur, for example, throughout Freud's *Introductory Lectures*:

> The era to which the dream-world takes us back is 'primitive' in a two-fold sense: in the first place, it means the early days of the *individual*—his childhood—and, secondly, in so far as each individual repeats in some abbreviated fashion during the course of childhood the whole course of the development of the human race, the reference is *phylogenetic*... It seems to me quite possible that all that today is narrated in the form of phantasy—seduction in childhood, stimulation of sexual excitement upon observation of parental coitus, the threat of castration— ... was in prehistoric periods of the human family a reality; and that the child in its phantasy simply fills out the gaps in its true individual experiences with true prehistoric experiences (Freud, 1922: 168, 311).

Freud's phyloculturalist preoccupations came, with the passage of time, to weigh as heavily as clinical considerations in the development of his 'metapsychological' theories—that is, theories of the overall (evolutionary—cultural) matrix within which psychoanalysis was to be situated. In *Totem and Taboo* (1953–74 [1912–13]) for example, Freud developed the hypothesis that the universality of the incest-taboo, the establishment of the Super-ego, and the centrality of the Oedipus complex, could all be traced to a 'primal scene'—itself partly based upon Darwin's speculations in *The Descent of Man* (1871)—wherein the ejected sons killed the Father/Ancestor.

Two points may be noted here. First, it was the biogenetic basis of Freud's phyloculturalist speculations that allowed him to dismiss criticisms (already current in his lifetime) that his theories were mere extrapolations from the culturally specific neuroses of his *haute bourgeoise* Viennese patients. Freud's assertion that 'the beginnings of religion, morals, society and art converge in the Oedipus complex' (1953–74 [1912–1913] vol. 13: 156) was logically connected to his belief, expressed in the 1915 edition of *Three Essays on the Theory of Sexuality*, that 'The barrier against incest is probably amongst the earliest historical acquisitions of mankind, and, like other moral taboos, has no doubt already been established in many persons by organic inheritance' (1953–74 vol. 7: 225); a remark which links Freud's 'metapsychology' directly to the evolutionary-developmental theories of J. M. Baldwin (see Chapter 4).

Second, Freud's interpretation of the 'totemistic' practices of 'primitive' societies may instructively be compared with that of Lévi-Strauss (1969 [1949]), who draws equally universal (and equally patriarchal) lessons from his structuralist reading of myths, but couched quite differently from Freud, in terms of an innate and species-specific 'grammar', the rules of which govern the exogamous 'exchange of women', rather than in terms of an essential-historical 'primal scene'.

Freud's propensity to weave myth out of myth is, of course, all too open to scientific criticism—though we would do well to remember that, at the time, the theoretical underpinning of the metapsychology was plausible enough. Still, if phyloculturalist speculation were *all* to Freudian theory, we would be justified in regarding it as yet another outgrowth of the rich foliage of nineteenth-century social biology. Aside from the clinical aspects of psychoanalysis, however, which do not directly concern us here, what compels continuing attention to the Freudian synthesis is its specific combination of methods, findings and hypotheses regarding the ontogenesis of symbolic and representational processes, their emotional investments, and their continuing consequences in adulthood.

The preceding quotations emphasised the correspondences which Freud postulated between the 'primal' scenes of, respectively, the infancy of the individual and the 'infancy' of the species. A complementary, and methodologically equally fundamental correspondence, between (ontogenetic) infant experience and adult manifestations of unconscious (or primary) processes was succinctly expressed in *The Interpretation of Dreams* (1954 [1900]: 546): 'a dream might be described as *a substitute for an infantile scene modified by being transferred on to a recent experience*'.

The hypothesis of the origin of dreams in the unconscious was the cornerstone of Freud's analysis, and the basis for his declaration that 'The interpretation of dreams is the royal road to a knowledge of the unconscious activities of the mind' (1954: 608). Freud's application of psychoanalytic principles to the analysis of dreams also marked the extension, after he had abandoned earlier theories in which *actual* childhood seductions played a central role in the genesis of neuroses, of the theory to apply to normal as well as pathological psychological phenomena. From the point of view of the ontogeny of symbolization, perhaps the most important aspect of the theory is the way in which it conceives dream-content and other psychic representatives, such as symptoms, as being the result of the operation of opposing dual principles.

These principles are most clearly set out in relation to the dream work. According to Freud, the wishes generated in the unconscious (UCs.), which is in effect the repository of repressed infantile memories and desires, 'cathect' or invest 'trains of thought' initiated in day-time consciousness. There then supervene *secondary* processes of censorship, which transform the dream-thoughts in such a way that the manifest dream content does not provoke the anxiety which would result from the direct expression in consciousness of the inadmissible wishes of what Freud would later call *das Es* (the Id). The semiotic aspect of Freudian theory thus consists in the specific analysis of the mechanisms of transformation effected by the secondary (pre-conscious) processes upon the psychic representatives of primary process (unconscious) material, in order to render such material admissible to consciousness. Thus, Freud wrote:

> The dream-thoughts and the dream-content are presented to us like two different versions of the same subject-matter in two different languages. Or, more properly, the dream-content seems like a transcript of the dream-thoughts into another mode of expression, whose characters and syntactic laws it is our business to discover by comparing the original and the translation (1954: 277).

The task of psychoanalysis, as Freud saw it, is simultaneously to elucidate the nature of the dream work, and relate it systematically to the theory of unconscious processes such as repression, identification, denial and so on. The two main processes constituting the dream work were termed by Freud *condensation* and *displacement*. Freud observed that very frequently a single element of dream content may be interpreted as simultaneously 'standing for' a variety of elements in the chains of different 'dream thoughts' representing different unconscious wishes. This 'overdetermination' of the unconscious meaning of elements of dream content lends to dreams their quality of 'condensation' as well as their apparently illogical character: 'Not only are the elements of a dream determined by the dream-thoughts many times over, but the individual dream-thoughts are represented in the dream by several elements' (1954: 284).

The process of condensation, resulting in the overdetermination of dream content, is frequently accomplished linguistically by punning, alliteration and other devices to which Freud drew attention in *The Psychopathology of Everyday Life* (1901); and the principal technique for investigating such processes psychoanalytically is the free association of verbal material. The work of condensation, however, is not the only means by which censorship is achieved: 'in the dream work a psychical force is operating which on the one hand strips the elements which have a high psychical value of their intensity, and on the other hand, *by means of overdetermination*, creates from elements of low psychical value new values, which afterwards find their way into the dream content. If that is so, a *transference and displacement of psychical intensities* occurs in the process of dream-formation, and it is as a result of these that the difference between the text of the dream content and that of the dream thoughts comes about' (1954: 308).

Although Freud stresses the significatory nature of condensation, and the energetic nature of displacement, the two aspects in fact coexist in both processes: the energetic aspect providing the motivation, and the semiotic or significatory aspct the mechanism, for the dream work and for symptom formation in general.[5] The applicability of linguistic-semiotic analyses of metaphoric (similarity) and metonymic (contiguity) relations in signifying systems[6] to the concepts of condensation and displacement has been noted by a number of writers, amongst them Jakobson (1956) and Lacan (1966), who maintains that 'the psychoanalysable symptom ... is supported by a structure identical to that of language' (p. 21).

It has frequently been suggested that the semiotic reading of Freud engaged in by Lacan and others[7] provides an alternative basis for a psychoanalytic theory 'cleansed' of biological determinism. This biologism has always rendered Freud's own interpretation of his work problematic for socialists and feminists wishing to harness psychoanalysis to the critique and transformation of the capitalist and/or patriarchal order (see, for example, Robinson, 1987). The perspective offered by the semiotic reading is the elimination of the dualism of the Freudian schema—a dualism between primary and secondary process, nature and culture, individual and society, affect and cognition—in favour of an analysis whose contradictions are played out on a non-biological plane—that of language, culture and the discursive order. Semio-psychoanalysis will then reveal and emancipate, not an underlying biological order repressed by culture, but the 'repressed' of discourse itself—the counter-discourse of the Other, played out beneath and beyond the phallocentric and logocentric structures of power and rationality (see Irigaray, 1978).

Two moments of socio-cultural critique are unified in the approach which I have sketched above: one which targets biological determinism as a recurrent and essential ideological support for the ruling order (see also Rose, 1982a, b; Rose *et al.*, 1984); and one, elaborated not only in contemporary feminist and post-structuralist critiques, but also in the Frankfurt School project and its heirs, which sees discursive rationality itself as being crucially implicated in the bureaucratic and scientistic modalites of contemporary forms of social domination. Although the force and importance of both of these critiques cannot be denied, both of them also contain extremely problematic elements. In the first case, anti-biolog*ism* is frequently elided (not, I should add, in the cases I have cited) into a denial of biology *tout court*, and a reliance on notions of the absolute autonomy of culture. In the second case, the arraigning of discursive rationality on the charge of 'logocentrism', threatens to collapse *all* coherent discourse into its instrumental-technical varieties.

This is not the place to address these issues in depth, but it might be suggested that the elimination of *every* aspect of duality from the Freudian schema threatens to undermine, not a 'classic', Cartesian dualism, but the very notion of a 'lived duality' of nature/culture which arguably distinguishes the Freudian analysis from other psychologies, and which lends it its dialectical character. Without this *necessary* duality, the Freudian approach would be subject to the same criticisms as have been levelled at Piaget's theory (see below). A (tentative) alternative suggestion for the reconstruction of Freudian psychoanalytic theory might be to focus, in a similar way as I do in relation to Vygotsky in this chapter, on the particular respects in which Freud's own psychobiological formulations are distorted by the Phylocultural Complex, and to reconstruct the interrelations in psychoanalysis between biology, semiosis and development in the light of more adequate formulations of all of these than were available to Freud himself.

Piaget: from solipsism to structure

Sigmund Freud and Jean Piaget (1896–1980) have probably been responsible more than any other individuals for shaping the discipline and practice of child and developmental psychology.

Amongst other similarities, they were both architects of grand synthetic theories which resist easy summary, and they were both profoundly influenced by evolutionary theories which, while current during their intellectual formations, now appear distinctly heterodox. Piaget, in particular, and even more explicitly than Freud, was a lifelong adherent and exponent of a neo-Lamarckian approach to evolution; and, while rejecting the simplicities of Haeckel's recapitulationism, he maintained a continuous dialogue in his developmental psychological theorizing with evolutionary biology. Piaget never in fact characterized himself as a psychologist, preferring the term *genetic epistemology* to designate the scientific discipline which he sought, if not to found (the phrase was apparently coined by J. M. Baldwin), at least to establish on a secure empirical footing.

As well as their common commitment to a genetic psychobiology embracing both ontogenetic and phylogenetic development, Freud and Piaget shared a common assumption regarding the psychic status of the newborn infant. This assumption may be summarized as a hypothesis of *original a-dualism* (Butterworth, 1981); that is, a lack of psychic differentiation between infant and environment, internal state and sensory surround. Freud called this state 'primary narcissism', a term emphasising the affective connotations of the hypothesis; Piaget dwelt rather upon the intellectual 'egocentrism' of the infant and the young child. For both theorists, however, 'His Majesty the Baby' is only gradually displaced from centre-stage, in a process involving cognitive and affective conflict.

Despite these similarities, however, the Freudian and Piagetian theories of the actual developmental process are quite different. For Piaget, the growth of subjectivity proceeds in an essentially harmonious manner, *via* the incorporation of *disequilibrated* conflictual states into higher-order, *equilibrated* 'structures d'ensembles'; whereas for Freud, the process is *inherently* contradictory, involving the gradual, and never wholly completed, subordination of primary to secondary process.

Again, for Piaget, original a-dualism is a logical initial postulate in a monistic, constructivist theory in which subject and object are co-constituted through action; in contrast, Freud's notion of original a-dualism is one of a primal condition which is interrupted and split by the emergent dualism of psychic processes, leading eventually to the establishment of the Ego and Super-ego. Consequently, the 'subjects' of Piagetian and Freudian theories are radically different entities. The Piagetian 'epistemic subject', situated within a progressive dialectical spiral of the growth of knowledge, is on a one-way track to higher, more equilibrated structures; though progress may be halted or retarded, it is irreversible. The 'progress' of the Freudian psychoanalytic subject, on the other hand, is altogether more precarious—it is continually prone to regression and disintegration; earlier infantile states, though repressed, threaten to return and haunt the present.

As Ingleby (1983) has noted, however, Piaget and Freud share a further presupposition, closely bound in both cases with the postulate of original a-dualism: that is, of the *a-sociality* of the infant. In Freud's case, consistently with the point of view I have outlined, this a-sociality is *dis*placed in adulthood, but persists in the Id; whereas, in the Piagetian scheme, the egocentrism of childhood is *re*placed by socialized cognition. Yet, and this is perhaps the most crucial aspect of Piagetian theory, the growth of egocentric thinking into de-centred, socialized thinking is accomplished *without* any significant role being accorded to social interaction or social structure *per se*.

The apparently a-social nature of Piagetian cognitive developmental theory has been the subject of criticism almost since Piaget's first formulations of it; and many of the critiques (including the early ones by Vygotsky and Wallon) have been couched in linguistic and semiotic, as well as psychological and philosophical, terms (see Venn and Walkerdine, 1978; Walkerdine and Sinha, 1978; Walkerdine, 1982). More recently, critique has been combined with attempts to reconstruct the Piagetian approach within a 'social cognitive' framework (see for examples and discussions of such approaches Butterworth and Light, 1982; Richardson and Sinha, 1987).

Other criticisms have concentrated on recent findings that infant perceptuo-cognitive competences appear to be more advanced than would be predicted by Piaget's theory, and have led to attempts to reconcile Piagetian with Gibsonian psychology (see Butterworth, 1981; and Chapter 4). Perhaps the most fundamental critiques of Piaget, however, are those which have suggested that his theoretical apparatus not only 'de-socializes' cognition, but tends also to identify it with the norms of rationality embedded in the instrumental-technical discourses of social regulation and domination (see for example Seltman and Seltman, 1985; Venn, 1984; Walkerdine, 1984). I do not have space here to address all of these issues; and I shall simply note that some (not all) of them are taken up in subsequent chapters. My principal focus in this section is on the exposition of Piaget's theory.

As is well known, Piaget's is a stage theory, in which the basic characteristic of any given stage of cognitive development is the manner in which actions (actual or internalized) are co-ordinated with respect to objects, and relations between objects, to which the actions are directed. Action is a more fundamental category than either subject or object, for through successive stages of the co-ordination of action, the object is *constructed* by the organism, in terms of categories such as causality, space, time and motion. Since these categories,[8] as Kant argued, form the necessary basis for logical reasoning, the growth of understanding on the part of the epistemic subject is to be construed as the growth of logic.

The logical (or logico-mathematical) structures constituting a given developmental stage are also to be understood as being *constructed*: logic, for Piaget, is both the measure and the product of the organism's developing interaction with the environment. This, in the baldest possible terms, is Piaget's theory, elaborated throughout a lifetime of empirical and theoretical work devoted to tracing the growth of logico-mathematical, operational thinking from the earliest stages of infant pre-intellectual action.

Piagetian theory conceives of human cognitive development as passing through three general stages or periods: the *sensori-motor*, lasting from birth until about 18–24 months, and itself divided into six sub-stages; the *pre-operational*, from the end of the sensori-motor period until the emergence of 'concrete operations' at around 6–8 years; and the *operational*, divided into the concrete and the formal operational stages, the latter emerging in late pre-adolescence or early adolescence. The central achievement of cognitive development at the end of the sensori-motor period consists in the *internalization* of actions as *representations*. Piaget is here his own best summarist:

> Sensori-motor intelligence is not yet ... operational in character, as the child's actions have not yet been internalised in the form of representations (thought). But in practice even this limited type of intelligence shows a tendency towards reversibility, which is already evidence of the construction of certain invariants.

> The most important of these invariants is that involved in the construction of the permanent object. An object can be said to attain a permanent character when it is recognized as continuing to exist beyond the limits of the perceptual field, when it is no longer felt, seen or heard, etc. At first, objects are never thought of as permanent; the infant gives up any attempt to find them as soon as they are hidden under or behind a screen ... Only towards the end of the first year does the object become permanent in its surrounding spatial field.
>
> The object's permanent character results from the organization of the spatial field, which is brought about by the child's movements. These co-ordinations presuppose that the child is able to return to his starting point (reversibility), and to change the direction of his movements (associativity), and hence they tend to take on the form of a 'group'. The construction of this first invariant is thus a resultant of reversibility in its initial phase. Sensori-motor space, in its development, attains an equilibrium by becoming organized by such a 'group of displacements' ... The permanent object is then an invariant constructed by means of such a group; and thus even at the sensori-motor stage one observes the dual tendency of intelligence towards reversibility and conservation (Piaget, 1977b [1952]: 456–7).

This quotation exemplifies many of the central themes of Piaget's work: the roots of knowledge in movement and action; the construction of reality by co-ordination of these actions; the subordination of perception to cognition; and the formalization of cognitive processes in terms of mathematics and logic (in particular, the logico-mathematical theory of sets, groups and operations). To continue, however: the stages of sensori-motor development may, at one level, be defined by the stages in the development of the object concept; however, another fundamental aspect of sensori-motor development is *the co-ordination of means and ends* in the organization of action schemata, and the co-ordination of schemata through reciprocal assimilation (Piaget, 1953 [1936]).

Whereas the *structural* aspect of Piaget's development theory is characterized in terms of logico-mathematical operations (or pre-logical 'operative actions'), ordered according to their degree of formal completion and flexible closure, the dynamic, *functional* aspect of development is characterized in terms of the dialectical interplay of *assimilation* and *accommodation*, of which the former is the more fundamental. Cognitive structures, at every stage of development, function as *assimilatory schemata*, analagous to the behavioural and physiological systems which enable the organism to assimilate food. All intelligence presupposes and depends upon assimilation, by which alone the structure of reality can be apprehended, and upon which perception itself depends. *Accommodation* however, is also implied by the very process of assimilation: every assimilatory act must be precisely 'fitted' (accommodated) to the particular object, event or logico-mathematical structure to which it is directed. Adaptation, for Piaget, consists of a process of *equilibration* between assimilation and accommodation.

The gropings of the infant towards means to re-enact pleasurable or interesting events (the event and the causative action being as yet undifferentiated) give rise to *circular reactions* in which elementary schemata are eventually assimilated to each other in parallel with the assimilation of object to schema. Eventually, this leads to the accommodatory ability to invent *new* means to achieve the same ends, through insight, which Piaget interprets as indicating the ability to co-ordinate schemata mentally (the internalization of action). Thus, the acquisition of a mature object concept, indexed by the ability to infer the position of a hidden object subjected to invisible displacements; the acquisition of elementary knowledge of spatial displacements independent of the actions of the subject; and the differentiation of means from ends, are all fundamental aspects of the emergence of *representation* at the end of the sensori-motor period.

Piaget (1953, 1962) defined representation—also termed the 'semiotic function'—in two (not always consistent) ways: first, in terms of the ability mentally to evoke absent realities, as implied by the notion of object permanence; and second, in terms of his own reading of semiotic theory. In this latter sense, Piaget distinguished between 'indices' or 'signals', which are already present in the early stages of sensori-motor development, and which directly 'trigger' assimilatory schemata; and 'symbols' which *represent* schemata, in virtue of their having a 'figurative' content different from the objects which they 'signify' *via* the schemata. Thus, 'symbolic signs' (including conventional linguistic signs) are closely related for Piaget to accommodation, inasmuch as the differentiation of the figurative signifier from the signified schema involves the internal differentiation of the schema itself (including means from ends).

Representational thinking provides the basis for developments in the pre-operational stage, including those involving language, deferred imitation, and symbolic play; all of which necessitate accommodatory adaptations to 'figurative' aspects of the environment. However, Piaget is quite clear that figurative knowledge depends upon, and is secondary to, operative (and ultimately logico-mathematical) knowledge. In other words, signification, for Piaget, is a *consequence* and not a *cause* of cognitive development. In particular, Piaget insists upon the primacy of action, and the cognitions deriving from the co-ordination of action, over linguistic knowledge: 'there exists a logic of co-ordination of actions. This logic is more profound than the logic attached to language and appears well before the logic of propositions in the strict sense' (cited in Furth, 1969: 122).

Piaget's denial of a formative developmental role to language and symbolization is of a part with his general belief that the figurative aspects of the environment in general are of only secondary importance: the fundamental basis of cognition being related to the internalization in operational thinking of the logico-mathematical structures which Piaget believed to be the necessary foundations of all knowledge, and thus the basis for genetic epistemology. For this reason, Piaget viewed all forms of linguistic determinism as incompatible with his constructivist approach.

Furthermore, the emergence of symbolization does not in itself bring about the development of reversible logical operations which Piaget takes to be the hallmark of concrete operational thinking; as evidenced, for example, in conservation, perspective-taking, and class-inclusion tasks. The attainment of representation and the concept of object permanence are only a first step in the passage from egocentric to de-centred thinking. Although the emergence of representation permits the development of language, the mastery of reversible thinking is not dependent upon, for example, the acquisition of certain lexical items, nor can it be hastened by training in verbal expression and comprehension. The same applies in principle to the transition from concrete to formal operations (co-ordinations of co-ordinations), although Piaget conceded that mastery of certain symbolic capacities may be a prerequisite for this final stage of cognitive development.

In concluding this section, I wish briefly to return to some further issues raised by recent critical discussions of Piaget's theory. The first concerns the relation, during sensori-motor development, between (a) the differentiation and co-ordination of means and ends; and (b) the construction of object permanence. Piaget considered these to be concurrent processes, both of which are necessary for the development of language and representational thinking. However, he acknowledged the possibility of a partial decoupling of the two during development, a phenomenon generally referred to in Piagetian theory as *décalage*. The question may then be posed as to *which* of these particular aspects of sensori-motor development is relevant and necessary to *which* particular aspects of language development.

Further, it may be noted that the significatory relations involved in both of these aspects of development, including their early stages, are of a *syntagmatic* or *metonymic* character, involving part–whole or contiguity relations. Those aspects of signification involving paradigmatic, metaphoric *substitution* relations are under-emphasized by Piaget; particularly inasmuch as they involve relations of iconic *similarity* This is of course consistent with Piaget's general lack of emphasis on the 'figurative'. The neglect of the figurative is symptomatic of the more general emphasis upon assimilation as more fundamental than accommodation, which underlies the whole of Piaget's theory, and in particular its 'internalist', a-social depiction of cognitive developmental processes. Within such an account, no *productive* role can be accorded to significatory processes. Nor, since the ultimate source of all signification is to be sought in action schemata, is there any room either for a theory of reference, as conceived by Frege, or for a theory of sign usage as a process of social and communicative exchange. Because Piaget attempts to account for all cognitive processes in terms of the intra-individual co-ordination of actions, there can be no formative role for signs themselves in cognition.

Vygotsky: language and the sociogenesis of reasoning

It has been said that the latter half of the nineteenth century witnessed three great dethronements of Man and Reason, in the theories of Darwin, Marx and Freud. Darwin not only evolutionized the 'Great Chain of Being', but, more radically, suggested that rationality itself was an adaptive *product* of nature, not a God-given faculty for the better understanding thereof. Freud pointed to the still-extant psychobiological inheritance of prehistoric dramas, and to the unconscious source in this prehistory of neurotic illnesses and other discontents of civilization. Marx, reversing the priority accorded since Descartes to the individual consciousness as the seat of Reason, stated that 'the human essence is no abstraction inherent in each single individual. In its reality it is the ensemble of the social relations' ('Theses on Feuerbach', Marx, 1975 [1845]: 423). Furthermore, social relations change, their transformations driven by the dialectic of class struggle and contradiction; thus, a materialist understanding of consciousness must also be of a historical and dialectical nature. Historical materialist analysis was therefore to provide the foundations both for a scientific analysis of class society, and for revolutionary socialist practice.

Lev Semenovich Vygotsky (1896–1934) was no less influenced by evolutionary theory than Freud and Piaget; but the particular forms of evolutionary thought which he took as his point of departure reflected the social, intellectual and ideological climate of post-revolutionary Russia in the 1920s, when he carried out most of his psychological work. Vygotsky's approach to psychology (often referred to as the 'historical-cultural approach'), like Bakhtin's approach to linguistics, with which it bears many affinities,[9] was therefore deeply marked by Marxist historical materalist theory. Vygotskian psychology is both *genetic*, or evolutionary-developmental; and *social*, emphasising the historical and cultural determination of forms of thought and symbolization.

This does not mean that Darwinian-evolutionary theories were unimportant in Vygotsky's thought. On the contrary, like most contemporary Marxists, he held evolutionary and historical materialisms to be complementary and mutually reinforcing doctrines. Moreover, Vygotsky strove to achieve a concrete synthesis of the two in his analysis of symbolization as tool-use, which provided the keystone for his psychology.

Darwinian theory first made an impact on pre-revolutionary Russian psychology through the work of Ivan Sechenov, a physiologist who pioneered the theory of reflexology which was later elaborated by Pavlov. As White (1983) has argued, Sechenov attempted to effect a synthesis of the experimental psychophysics of Helmholtz and Fechner with the evolutionary theories of Darwin, as popularly propounded by Herbert Spencer.[10] Sechenov's book *Reflexes of the Brain* (1935 [1863]), which, as was intended by Marx for *Capital*, bore a dedication to Darwin, was considered by the Czarist censor to be sufficiently subversive to warrant its brief suppression. The revolutionary import of Sechenov's work lay in his contention that a materialist, reflexological psychology was indifferently applicable to the study of both human and animal behaviour. In this and later works Sechenov also made frequent reference to stages of ontogenetic development, by means of which the child progresses from 'concrete', sensorial, to 'abstract', representational reasoning: a mode of argument which links Sechenov not only back to Spencer, but forward to Piaget and Vygotsky.

In the course of the 1920s, the materialist reflexology of Sechenov and Pavlov was consecrated as official Soviet psychology, in a repudiation of idealist and introspectionist psychologies.[11] Vygotsky, however, noted that both introspectionism and reflexology shared a common commitment with behaviourism to a stimulus–response model in which 'higher mental processes' were either ignored (as in reflexology), or deemed unamenable to experimental study (as in introspectionism). The active contribution of the subject, in both cases, was neglected. Thus, while Vygotsky welcomed the objective and scientific method underlying reflexological psychology, he criticised its neglect of consciousness and subjectivity. What distinguished Vygotsky's position from, for example, that of the Gestalt psychologists, was his insistence that the material substrate of consciousness was to be sought partly in the socially and historically determinate conditions of its acquisition and development.:

> Within a general process of development, two qualitatively different lines of development, differing in origin, can be distinguished: the elementary processes, which are of biological origin, on the one hand, and the higher psychological functions, of sociocultural origin, on the other. *The history of child behaviour is born from the interweaving of these two lines.* The history of the development of the higher psychological functions is impossible without a study of their prehistory, their biological roots, and their organic disposition. The developmental roots of two fundamental, cultural forms of behaviour arise during infancy: the use of *tools* and human *speech* (Vygotsky, 1978 [1930]: 46).

Speech and symbolization—sign-usage—stand at the centre of Vygotsky's theory of the sociogenesis of reasoning. Pavlov had already recognized the importance of speech as a 'second signal system', a *mediator* between stimulus and response underlying neo-cortical functioning. Vygotsky drew on this analysis of mediating function in comparing (but not identifying) sign and tool use, and in analysing their co-functioning.

> The basic analogy between sign and tool rests on the mediating function that characterises each of them ... [However] a most essential difference between sign and tool, and the basis for the real divergence of the two lines [of their development], is the different ways that they orient human behaviour. The tool's function is to serve as the conductor of human influence on the object of activity; it is *externally* oriented; it must lead to changes in objects. It is a means by which human activity is aimed at mastering ... nature. The sign, on the other hand, changes nothing in the object of a psychological operation. It is a means of internal activity aimed at mastering oneself; the sign is *internally* oriented ... The mastering of nature and the mastering of behaviour are mutually linked, just as Man's alteration of nature alters Man's own nature. In phylogenesis we can reconstruct this link through fragmentary ... evidence, while in ontogenesis we can trace it experimentally ... The use of artificial means, the transition to mediated activity, fundamentally changes all psychological operations, just as the use of tools limitlessly broadens the range of activities within which the new psychological functions may operate. In this context, we may use the term *higher* psychological function, or *higher behaviour*, as referring to the combination of tool and sign in psychological activity (Vygotsky, 1978: 54–55).

The joint mediation of activity, by sign and tool, enables the child to gain increasing control over his or her own actions, through first, the use of an external sign as a 'tool' of consciousness (Vygotsky uses the example of tying a knot in one's handkerchief as an *aide-mémoire*); and, second, through the later *internalization* of such sign-usage in the form of 'higher functions' such as voluntarily directed memory and attention. The notion of 'internalization', together with that of mediation, is fundamental to Vygotsky's historical-cultural approach:

> The internalization of cultural forms of behaviour involves the reconstruction of psychological activity on the basis of sign operations. Psychological processes as they appear in animals actually cease to exist; they are incorporated into this system of behaviour and are culturally reconstituted and developed to form a new psychological entity ... The internalization of socially rooted and historically developed activities is the distinguishing feature of human psychology (Vygotsky, 1978: 57).

Vygotsky also laid considerable stress upon the role of actual social interactions in the internalization of higher functions, an emphasis which, as already noted, has led many recent writers directly to compare his position with that of Mead. A much-quoted dictum of Vygotsky would certainly support such a comparison:

> Every function in the child's cultural development appears twice: first, on the social level, and later, on the individual level; first, *between* people (interpsychological), and then *inside* the child (intrapsychological) ... All the higher functions originate as actual relations between human individuals (Vygotsky, 1978: 57).

As Holzman (1985) has noted, however, comparisons between Mead and Vygotsky should not conceal the differences in their approaches. Whereas Mead took communication and interaction to be *constitutive* of 'social process', for Vygotsky the *historical* process rather determined the available repertoire of signifying and other psychological activities, and thus, ultimately, the actual course of development. The historical materialist basis of Vygotsky's developmental psychological theory has subsequently been developed by Soviet psychologists such as Luria and Leontiev, and I shall discuss the role of history in Vygotsky's theory more fully in the next section of this chapter. Meanwhile, no summary of Vygotsky's theory is complete without reference to his notion of 'inner speech', and the critique he offered of Piaget's conception of infantile egocentrism.

Piaget had introduced the notion of 'egocentric speech' in his first book, *The Language and Thought of the Child* (1926), in which he observed that many of the utterances of pre-school children appeared to be self- rather than other-directed. This he took to be symptomatic of the general intellectual egocentrism of the child. As Vygotsky noted, 'the development of thought is, to Piaget, a story of the gradual socialization of deeply private, personal, autistic mental states. Even social speech is represented as following, not preceding, egocentric speech' (Vygotsky, 1986: 34). Vygotsky, on the other hand, both emphasized the functional significance of 'egocentric' speech, and interpreted its developmental status quite differently, in terms of his concept of internalization. This is how Vygotsky summarized his viewpoint:

> The primary function of speech, in both children and adults, is communication, social contact. The earliest speech of the child is therefore essentially social. At first it is global and multifunctional: later its functions become differentiated. At a certain age the social speech of the child is quite sharply divided into egocentric and

communicative speech ... Egocentric speech emerges when the child transfers social, collaborative forms of behaviour to the sphere of inner-personal [intrapsychological] functions ... Egocentric speech as a separate linguistic form is the highly important genetic link in the transition from vocal to inner speech, an intermediate stage between the differentiation of the functions of vocal speech and the final transformation of one part of vocal speech into inner speech ... Thus our schema of development—first social, then egocentric, then inner speech—contrasts both with the traditional behaviourist schema—vocal speech, whisper, inner speech—and with Piaget's sequence—from non-verbal autistic thought through egocentric thought and speech to socialized speech and logical thinking ... In our conception, the true direction of the development of thinking is not from the individual to the social, but from the social to the individual (Vygotsky, 1986: 35–36).

According to Vygotsky, inner speech represents the fusion of the hitherto-separate developmental paths of 'prelinguistic thought' and 'pre-intellectual speech'; giving rise to consciously directed thought processes and, ultimately, logical and scientific thinking. Thus, the notion of inner speech was closely tied to another Vygotskian theme, that of the *directive* function of speech and language in cognitive activity. This aspect of Vygotsky's thought was taken up by his co-workers (notably A. R. Luria) after Vygotsky's own early death from tuberculosis at the age of 37. It also reflected a general emphasis in Vygotsky's work on the *interdependence* of language and thought, a position further distinguishing Vygotskian from Piagetian developmental theory. For Piaget, as we have seen, the 'figurative' aspects of symbol acquisition and use, including language, were subordinate to operative development. For Vygotsky, on the contrary, the development of language itself permitted the differentiation and elaboration of thought.

The relation of thought to word is not a thing but a process, a continual movement back and forth from thought to word and from word to thought. In that process the relation of thought to word undergoes changes which themselves may be regarded as development in the functional sense. Thought is not merely expressed in words; it comes into existence through them (Vygotsky, 1986: 218).

Although Piaget and Vygotsky differed considerably in their developmental theories—over the roles which they assigned, for example, to social interaction and language in cognitive development—there was also much which they shared. Both theorists stressed the lack of 'logical' or 'scientific' concepts in young children; both stressed the differences, rather than similarities, between the representational and symbolic capacities of children and adults; and both proposed a form of constructivism as an alternative to both empiricism and nativism.

Furthermore, Piaget and Vygotsky shared with Freud a common intellectual heritage, in the form of the Phylocultural Complex. In the case of Freud, recapitulationist and parallelistic notions have been shown to have played a central role in his theoretical development. The early works of Piaget and Vygotsky, too, as evidenced by the quotations at the end of Chapter 2, were marked by appeals to parallelism, in the form of child-primitive comparisons, derived at least in part from the writings of the anthropologist Lévy-Bruhl (e.g. Lévy-Bruhl, 1918). Neither Piaget or Vygotsky, however, were adherents of recapitulationism in the sense of Haeckel's biogenetic law, and both attempted subsequently to deploy alternative concepts in order to produce a more satisfactory and less reductionist account of the relations between biological, social and cognitive development. In the case of Piaget, the crucial concept was that of *epigenesis*, which forms a central focus for Chapter 4.

As far as Vygotsky is concerned, Scribner (1985) argues that his use of child-primitive comparisons was motivated not by the more usual notion of parallelism, but by his insistence upon the central role of *history* in both ontogenesis and phylogenesis. In particular, she suggests that the concept of 'general history', understood as the emergence of 'specifically cultural' (though not culture-specific) forms of behaviour and their regulation, provided the mediating concept in Vygotsky's attempt to 'achieve a synthesis in psychology between "nature" and "culture"' (p. 123). In the final section of this chapter, I provide an evaluation and critique of Vygotsky's general use of temporal and historical categories in his process of theory construction, suggesting an alternative model, consistent with his general approach, but differing in its specification of developmental mechanisms.

Ontogeny and phylogeny: beyond the paleomorphic metaphor

For any evolutionary and developmental theory, a fundamental problem concerns how time—or rather temporalities in the plural, since more than one time, or *durée*, is involved—may best be conceptualized (see Gould, 1987). Darwinian evolutionary theory revolutionized previous European conceptions of time. Darwin's theory went further than simply to valorize the static 'Great Chain of Being' with an evolutionary, temporal dimension (Lovejoy, 1936). It transformed time itself, splitting and multiplying the scales by which it might be measured. No longer could time be seen as originating at the moment of creation; and, if there was no longer a single First Cause, there was also no longer a single First Event. With the demise of the reciprocal notions of Origin and Finality, it fell to evolutionary theorists to attempt some account of the relations between the time scales—earth historical, human historical, and ontogenetic—which might also illuminate the course of human evolution itself.

One of the earliest and most widely accepted such attempts, as we have seen, was recapitulationism. As Reynolds (1981) has argued, however, recapitulationism and the 'Phylocultural Complex' can best be seen, not so much as real transformation of previous ideas of time, as an assimilation of Darwin's discoveries (an assimilation of which Darwin himself was more than once guilty) to existing static ideas of a *scala naturae*—with, of course, European Man at its summit. Victorian evolutionary theory was typological, rather than dynamic and historical, and its use of the comparative method largely consisted in 'demonstrations' of primitivity and difference, rather than continuity.

The discrediting of the simplicities of recapitulationalism threw the relations between ontogeny and phylogeny once more into question. Piaget, for example, although no follower of Haeckel, wrote at length on the question of evolution and individual development (e.g. Piaget, 1979), and was a highly original evolutionary thinker, seeking to provide a principled scientific basis for his quasi-Lamarckian views in the concept of epigenesis. Paradoxically, however, in view of his self-designation as a genetic epistemologist, Piaget neglected time and temporality in most of his writings. Time, for Piaget, is reducible to the invariances of sequence which are inbuilt in the relations between the logico-mathematical structures characterising the developmental stages of the cognitive or epistemic subject. Unlike Vygotsky, Piaget did not address the issue of the time scale, or *durée*, specific to human societies—that of history.

The problem of time has a similarly contradictory status, encompassing both recognition and neglect, in twentieth-century linguistic theory. On the one hand, it was central to the structuralist project founded in the early years of the century by Saussure; the time-referring concepts of synchrony and diachrony underlie the langue-parole distinction which he took to constitute the very possibility of a science of language. On the other hand, as writers from the Prague Linguistic Circle (see Jakobson and Tynjanov, 1985 [1928]) to the present day (Bailey and Harris, 1985) have noted, the structuralist focus on synchrony, langue, and abstract universal competence, has tended to obscure the importance both of language change and development, and of the particularities of socio-cultural process.

The structuralist tradition, following in the footsteps of the Enlightenment philosophers, has viewed language as that which sets humanity off from other species, enabling the development of culture and rationality.[12] Inevitably, such an approach tends, first, to treat 'language' and 'culture' as universal terms, and, second, to view the problem of ontogenesis in terms of a transition of the infant from an animal/biological to a human/social state. This basic notion of 'socialization' is common to such diverse thinkers as Freud, Piaget and Vygotsky, whatever their other differences. For all of these theorists, language/culture/symbolization stands as the *sine qua non* of the human condition itself—and thus of human history; and history/society/culture as the *sine qua non* of the socialization of the individual.

If the dualistic opposition between 'nature' and 'culture' can be seen as having its roots in the work of the eighteenth-century *philosophes*, the idea that time itself is 'layered' is of more recent origin, being principally inspired by Darwin's revolutionary reassessment of the implications of the geological and palaeontological investigations of Lyell and others. The importance of palaeontology in establishing Darwinian evolutionary theory led to the appropriation by many evolutionists of palaeontology as a fundamental metaphor, not just for changes *within* a *durée*, but for relations *between durées*. What underlies this 'palaeomorphic' model, or metaphor, is a hierarchy of temporalities: evolutionary-biological time (itself embedded within geological time) is presupposed by social-historical time, which is presupposed by ontogenetic-developmental time.

Within the terms of the model, one may postulate parallelisms of sequence or mechanism, as did Freud and Piaget with the respective concepts of recapitulation and epigenesis; and one can argue over the relative influences of more 'fundamental' levels over more 'superficial' ones, as in Vygotsky's insistence upon the equivalent importance of the social and the biological in individual development. What remains unchallenged, however, is the basic assumption that time itself is stratified, each 'layer'—geological, evolutionary, historical, individual—being sedimented upon the previous one (see Fig. 3.1). This palaeomorphic metaphor was explicitly stated in relationship to behavioural development by Vygotsky, in the following terms:

One of the most fruitful theoretical ideas genetic psychology has adopted is that the structure of behavioural development to some degree resembles the geological structure of the earth's core. Research has established the presence of genetically differentiated layers in human behaviour. In this sense the geology of human behaviour is undoubtedly a reflection of "geological" descent and brain development. If we turn to the history of brain development, we see what Kretschmer calls the law of stratification in the history of development ... lower centres are retained as subordinated structures in the development of higher ones

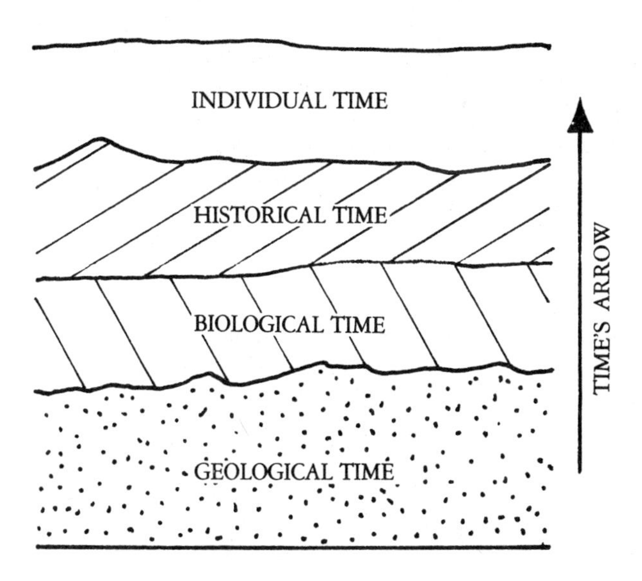

Figure 3.1: The palaeomorphic metaphor: time as geology

> and ... brain development proceeds in accordance with the laws of stratification, or construction of new levels on old ones ... Instinct is not destroyed, but "copied" in conditioned reflexes, as a function of the ancient brain which is now to be found in the new one. Similarly, the conditioned reflex is "copied" in intellectual action ... the behaviour of the modern, cultural, adult human can be understood only "geologically", since various genetic layers, which reflect all the stages through which humans have travelled in their psychological development, are reflected in it (Vygotsky, 1981: 155–156).

It is in this context that Vygotsky proposes that the development of 'higher mental functions' be viewed as representing a 'completely new level of development', in which 'any higher mental function was [once] external, because it was social, at some point before becoming an internal, truly mental function. It was first a social relation between two people.' Vygotsky goes on to restate his 'general genetic law of cultural development: any function in the child's cultural development appears twice, or on two planes. First it appears on the social plane, and then on the psychological plane ... higher functions are not developed in biology [phylogenesis]. Rather, the very mechanism underlying higher mental functions is a copy from social interaction; all higher mental functions are internalized social relationships' (Vygotsky, 1981: 162–164).

This quotation emphasizes the importance of the concept of internalization as the fundamental mechanism by means of which 'general history' enters into ontogenetic processes. This mechanism, however, suffers from a serious logical problem. If the individual cognitive subject is seen as being an internalized product of social life and organization, and not a product of biology, then what is the nature of the subject (or proto-subject) which is initially responsible for the act(s) of internalization? To say that this is itself biological is simply to push the problem down a level, for the capacity to become 'fully human' is also a uniquely human characteristic. An equivalent dilemma is faced by theories which attempt to solve the problem 'from the other side'—by appeal to the intersubjective structuration of early actions and interactions (Trevarthen and Hubley, 1978; Lock, 1980). Granted the importance of attributions of intentionality to infants by adults, and of adult 'scaffolding' (Bruner, 1983) of early interactions, these again presuppose a subject capable of 'copying' the interpersonal attributions and transactions onto the 'inner-psychological' plane.

Such considerations lead to the conclusion that, despite its interactionist and dialectical impulses, the Vygotskian theory of internalization reproduces in its internal logic the very divisions between the natural and the cultural, and the individual and the social, which it strives to overcome. Furthermore, the palaeomorphic model of layering and sedimentation of neurological and psychological function, corresponding to the layering of temporalities (ontogenetic upon socio-cultural upon phylogenetic), reproduces the classical division between 'instinctual' and 'learned' behaviour; leaving the Vygotskian account open to the criticism that it does not really break with the behaviourist and empiricist paradigm (Fodor, 1972).

The embedding of the theory of internalization within the palaeomorphic metaphor both generates the logical problem at the heart of the theory—a logical problem which, of course, underlies any account based upon the 'socialization of the non-social'—and undermines the attempt of Vygotsky to carry through an epistemological break with previous theories, by recouping his concept of learning to the familiar empiricist notion of representation as mental–neural 'copying' (albeit the copying is now *mediated* rather than direct). A non-empiricist recasting of Vygotskian theory therefore requires a critical re-examination of the palaeomorphic metaphor itself, and its replacement by a more adequate account.

There is, at one level, an obvious truth in the palaeomorphic metaphor. Both individual development and social life do indeed depend upon an evolutionary biological process which for the greatest part pre-dates the emergence of our species, and whose *durée* is of an entirely different order from that of even the longest historical span.[13] We may, for example, suppose that if a newborn Neolithic infant were somehow transposed with one from a present-day delivery room, the subsequent physical, psychological and social development of the changeling would be commensurable with those of its new-found peers. Thus, *both* what is constant in human development *and* what is variable across cultures are equally supported by a genetically transmitted biological 'core'. This much is unquestionable, as is the relevance of genetic transmission for variation between individuals within cultures.

What *is* open to question, however, is the assumption that the biological 'core' of individual development—the organismic 'support' necessary for the growth of human subjectivity—is itself a product *solely* of biological evolution. This proposition, stated negatively—'biology is not a product solely of biology'—has a paradoxical ring to it, and so is perhaps better stated in the following terms, as a positive hypothesis: 'the biology of human development is a product of the interaction of biological and cultural evolution at the specific site of ontogenesis'.

What I am proposing, then, is that rather than seeing cultural evolution as 'taking off' from a terminal point of biological evolution, we should rather see evolutionary biological processes as having been, as it were, 'captured' by an emergent cultural process, with ontogenetic processes—especially those involving representation, symbolization and communication—as a crucial catalyst and product of the co-evolution of culture and biology.[14] As Vygotsky, and many others, including his colleagues Luria and Leontiev, have emphasized, the emergence of human symbolic and representational capacities, in both ontogenesis and phylogenesis, is intimately connected with the emergence of cultural transmission and tool use. However, merely to state this is in a way to state the obvious; and, since neither tool use nor cultural transmission are unique to the human species, without further specification such a theory lacks explanatory power.

The theoretical position which I wish to advocate begins with a re-assessment of the notion of representation, with particular emphasis on the materiality of representation, and the representational environment. The human environment, for any concrete individual, has always-and-already been intentionally shaped by previous generations of human agents into a material culture. Of course, biology—and in particular neurobiology—is an indispensable precondition or support for symbolic representation. However, a crucial aspect of the human enviroment is that it involves representational systems which extend *beyond the boundaries of the individual organism*. Representation, on this account, is not simply a distinctively human cognitive and cultural capacity, but a constitutive property of the material surround into which the human infant is born, and which, like all environments, supports and constrains the activities of the developing organism.

Consider, for example, tool use. Tools are artefacts functioning as adaptive extensions of organismic movements, and as such the structure of a tool may be seen as materially representative both of organismic structure (and the behaviours that structure permits), and of the local environment within which the tool functions. Tools therefore both *support* and *constrain*, *via* a material representational structure, adaptive organismic functioning, and, moreover, do so in the case of human societies within a cultural context of co-operative praxis and the division of labour—features of context which equally find a material representation in the structural properties of the means of production.

Again, consider language. Linguistic structures, at a phonological and syntactic level, are representational systems insofar as they are the means for articulating messages governing and sustaining interpersonal interaction. These structures, however, by virtue of their *exteriority* and *objectivity*—i.e. their independence (at a 'virtual' level) of the instantiation of *particular* messages—*constrain* communicative practices to the structure itself: language, too, functions as both a support for, and a constraint upon, practice. It is this dialectic between support and constraint which precisely characterizes the *materiality of representation.*

As Leontiev (1981 [1931–1960]: 134) puts it: 'Before the individual entering upon life is not Heidegger's "nothing", but the objective world transformed by the activity of generations.' To this we might add that the 'objects' in the world encountered by the child are neither *simply* Newtonian particles obeying universal physical laws, nor are the actions in which they are embedded *simply* exemplifications of abstract logico-mathematical operations, as a reading of Piaget might suggest.

In the first place, as many authors have noted (Trevarthen and Hubley, 1978), the infant early encounters 'social' or 'personal' objects—simply, other people—whose qualities are different from those of impersonal objects. In the second place, however, *impersonal* objects encountered by the infant are *also* social, in a different sense, in that they are encountered within a context of particular social practices and social relations. Furthermore, in the case of artefacts in general, as for tools in particular, the context-of-use of the object commonly achieves, as I have emphasized, representational status in the structure of the object itself.

More artefacts are designed to fulfil a certain purpose, which we may designate the *canonical* (or socially standard) function of the object, and this canonical function (containment, cutting etc.) in part determines the *form* of the object (possession of a cavity or blade etc.). Such relations between functions and forms may be termed, in general, design rules, and any artefact may be seen as a material representation of a subset of the design rules current in a culture. Such rules are not strictly deterministic, since although they are constrained by physical laws, such constraint is weak. A particular function may be subserved by many different forms: for example, the function 'key' may be subserved by a hand- or machine-engineered implement, a mechanical combination-lock sequence, or a sequence of characters entered at an electronic keyboard. In general, we may say that the range of available instantiations of a specified function will increase with increasing knowledge and technique, as will the range of actually available functions. Furthermore, certain instantiations may become obsolete, so that design rules are subject to historical change.

Nevertheless, although design rules are underdetermined by function, they are not arbitrary, since they embody constraints of nature and of social practice, habit and sometimes aesthetic value. Furthermore, a certain core of design rules, governing the use of artefacts common to all cultures—implements for containing, supporting, cutting, pounding, tying etc.—remain relatively speaking invariant across space and time, and it is this subset of invariant design rules, correlating canonical (that is, socially standard) functions and canonical forms, which may be designated as *canonical function-form relations*. My proposal is that such canonical relations act as a material, representational core to object usage, and possess a privileged status in both ontogenetic and phylogenetic cognitive and symbolic development.

As I shall show in more detail in Chapter 5, the child's early accommodations to, and assimilations of, objects are significantly determined by the social and intentional shaping of these objects to represent, in their structural properties, canonical relations. In this respect, the object provides an external epigenetic pathway (Waddington, 1977) for what Leontiev (1981) calls the 'appropriation' of knowledge and culture by the developing human organism.

The canalization of cognitive growth by the object is reinforced by the tuition provided by adult co-participants in joint and routine interpersonal and communicative activities. To quote Leontiev (1981: 135) again, 'the individual, the child, is not simply thrown into the world; it is introduced into this world by the people around it, and they guide it in that world'. The extraordinarily high degree of senstivity displayed by the human infant and young child to the cues provided, directly through interpersonal communication channels, and indirectly through context-setting, by adult interlocutors has been documented by many developmental psychologists in recent years, as has the dyadic structure of early tutorial and pedagogical relationships (Wood, Bruner and Ross, 1976).

Representational and symbolic development do not, however, end with the consolidation of knowledge of canonical relations: rather, this consolidation forms the basis for subsequent developments in which the child systematically violates canonical relations, in symbolic and substitutional pretend play routines, and in constructional goal-oriented activities. The ability to de-couple canonical function–form relations is crucial to language development, particularly the acquisition of syntax; and in general to the generation of novel means–ends relations. This fundamentally creative or innovative capacity is as much, or indeed more, a specifically human cognitive capacity as the ability to produce enduring tools and artefacts; furthermore, the ability to go 'beyond the information given' (Bruner, 1974) is combined, in the human species, with the emergence of constructional activities which involve the co-ordination of different but complementary roles in the social organization of the task (Reynolds, 1982).

Thus, if the grasp or appropriation of canonical relations may be seen as a fundamental mechanism for the social transmission of the core invariants of culture, symbolic play and co-operative action provide ontogenetic mechanisms for the generation of complex substitutional and combinatorial systems, and ultimately for the appropriation of the generative and combinatorial rules (language, kinship systems etc.) which are specific to particular cultures. What implications do these ontogenetic processes hold for possible mechanisms of the phylogenetic evolution of cognition and symbolization?

My proposal is that the emergence of culture and symbolization, in its human form, was made possible by the evolution of infancy itself as a specific niche, whose adaptive parameters were set by the prior emergence of hominid artefact manufacture and use. Ontogenesis, then, was the specific evolutionary adaptation permitting the inter-generational transmission through appropriation of the earliest cultural artefacts, those instantiating canonical relations. According to this hypothesis, it was not tool use *per se* which was responsible for the emergence of human cognitive and symbolic capacities, but the enduring products of such tool use in the form of artefacts (including tools themselves).

As Leontiev has emphasized, the process of appropriation is not simply one of adaptation to a given reality, but of the active mastery of the transformative procedures which tools and artefacts represent and potentiate; furthermore, the process of appropriation is fundamentally social, in two distinct but related senses. First, in that the procedures mastered and reproduced through appropriation are inseparable from the social order as a whole, and second in that the guidance of adults is an intrinsic part of the developmental process. Therefore, it is not the case that infancy is adapted to tool and artefact use *per se*, and I am not arguing that tool use is 'innate' in the conventional sense.

Rather, the biological adaptations necessary for human evolution were selected by the requirements of appropriation as an ontogenetic mechanism. Thus, the evolution of infancy was the biological mechanism through which the potential for inter-generational cultural transmission created by tool use was optimized, and appropriation was the specific social and ontogenetic process in response to which the niche of infancy evolved. In this respect, the evolution of infancy was not so much a terminal point of biological evolution, but the crucial inaugural moment in the socialization of biology (Riley, 1978).

Such an evolutionary process, I suggest, would lead to a rapidly expanding endogenous spiral of change and adaptation. Those aspects of infant psychology facilitating appropriation processes, such as social-contextual sensitivity, representational and symbolic control of sensori-motor processes, and 'dedicated' perceptuo-motor mechanisms for processing speech and gesture, are also those which are most clearly implicated in the crucial features of a specifically human form of life: symbolic and representational systems, technological innovation, complex classificatory systems and so forth.

The hypothesis that I have presented brings ontogenetic processes once again to the fore in theories of human symbolic, cognitive and cultural evolution, while departing significantly from the assumptions of the 'palaeomorphic metaphor'. Specifically, this hypothesis suggests that certain features of the biology of human infancy and early childhood—those associated with appropriation procedures—are evolutionary adaptations which postdate the emergence of artefact use, but predate (and potentiate) the emergence of complex categorization and social-technological capacities. This theoretical account also leaves room for the evolution of certain 'modular' human capacities, including those involved with processing language and gesture, as part of the adaptive response to the niche of infancy.[15] However, this is not the same as saying that 'linguistic rules' are innate, any more than to suggest that innate pre-adaptation to the appropriation of canonical relations is the same as saying that such relational rules themselves are innate. The innateness of the rules (of either grammar or artefact use) cannot be excluded by, but nor is it implied by, the theory I have presented.

Notes

1. The idea that stages in individual development parallel 'grades of organization' in the animal kingdom in fact predated Darwinian evolutionary theory, having been proposed—according to Thomson (1920)—by Meckel, von Baer and Agassiz, amongst others.
2. It should be noted that the ideological-administrative basis of child psychology, from its inception, was not restricted to the regulation of non-European populations. As has convincingly been argued by Ehrenreich and English, 1978; Riley, 1983; Urwin, 1985; and Walkerdine, 1985 it has also been inextricably implicated in practices and discourses of the moral regulation of metropolitan 'subject populations'; with the working-class family (particularly mother and child) as its principal site of intervention.
3. Sulloway comments that Freud 'abandoned one reductionist goal in science—a mechanical-physiological one—and, without being fully happy about it, adopted another, and more evolutionary one, in its stead' (p. 126).
4. See, however, for a contemporary re-evaluation of Lamarckism, Ho and Saunders (1984).
5. For an extended discussion of 'informational' and 'energetic' concepts in Freud's theory of psychic process, and particularly in relation to the 'Project for a Scientific Psychology', see Pribram and Gill (1976).
6. The metaphor–metonymy distinction is roughly equivalent to that between paradigmatic and syntagmatic relations. cf. chs. 1 and 5.
7. I shall ignore, for reasons of space, both the residual biologism of the

Lacanian account, and the role of that residual biologism in motivating the fierce rejection by Lacan of previous 'culturalist' readings of psychoanalytic theory, such as those of Fromm and Horney.

8. Kant actually distinguished space and time as 'forms of intuition' from 'categories of the understanding' such as causality; however, this has no immediate bearing on the discussion.
9. According to Clark and Holquist (1984) there is, however, no evidence that Vygotsky and Bakhtin were acquainted with each other's work. Wertsch (1985) suggests that a common source of influence was Yakubinsky, 1923.
10. See also Cole and Scribner's introduction to Vygotsky, 1978.
11. See Kozulin's introduction to Vygotsky, 1986.
12. As Reynolds (1981: 13) notes, the ultimate collapse of Victorian phyloculturalist theories, with their assimilation of evolutionary concepts to older ideas of a *scala naturae*, was accompanied in anthropological and social theory by a return to a dualistic distinction between culture and nature, with 'an evolutionary interpretation of the human body and ... a historicist theory of human action'. It was, of course, precisely this dualism which Vygotsky in particular sought to transcend.
13. In this sense, Schilcher and Tennant (1984: 114) are absolutely correct in asserting that 'We must constantly bear in mind that it is the biological level that subtends the cultural, and not the other way round.'
14. Similar suggestions have been made by Geertz, 1973 and Sahlins, 1976.
15. The issue of modularity and 'computational specificity' will be taken up further in the next chapter.

4 Language, Mind and Nature

What in a particular case of motor functions in organisms appears to be (1) the modelling of future requirements in terms of a problem of action, and (2) the realization of an integrated program of this action by the conquest of external obstacles and by an active struggle for the result, turns out to be a manifestion of the general principle of activity running through the whole of biology.

N. A. Bernstein (1967: 175)

Matter is not the *Ground* of Form, but the unity of Ground and Grounded.

G. W. F. Hegel[1]

This is a gift that I have, simple, simple; a foolish extravagant spirit, full of forms, figures, shapes, objects, ideas, apprehensions, motions, revolutions. These are begot in the ventricle of memory, nourish'd in the womb of pia mater, and delivered upon the mellowing of occasion.

William Shakespeare[2]

The modern syntheses

In this chapter I examine traditional and contemporary approaches to psychobiology and evolution, with particular reference to their implications for cognitive theories. I develop an argument for one particular approach—that is, the epigenetic one—by way of a critical examination of alternative and opposing paradigms in both biological and cognitive sciences, which may be regarded as the dominant 'modern syntheses' within these two fields. I argue that these two modern syntheses—namely, neo-Darwinism and neo-rationalism—are not only, taken individually, inadequate accounts, but that since the demise of the 'Classical synthesis' of the early twentieth century (the combination of reflexology and Darwinism), they do not offer, taken together, the promise of a coherent and unified psychobiological account. This analysis prepares the way for an examination of how the epigenetic approach—which accepts many of the premises of neo-Darwinism and many of the arguments of neo-rationalism, while differing substantially from both—does offer precisely such a unified and coherent account, which is moreover compatible with the analysis of representation and its development presented in other chapters, and with recent research in the neurosciences.

The term neo-Darwinism is intended, in what follows, to refer to that approach to evolution which is predicated upon the so-called 'central dogma', owing originally to the work of Weismann (1972) [1885]: that the only source of evolutionary change is natural (and sexual) selection of randomly generated variations in genetic material. In contemporary versions of neo-Darwinism, this is supplemented by the findings of modern population genetics, and the model theoretically-derived mechanisms advanced by sociobiologists, such as kin selection, differential reproductive strategies etc. (Wilson, 1975).

It is important, though, to emphasize that epigeneticism, as a general psychobiological approach, is in opposition to neo-Darwinism *only* inasmuch as it questions the *exclusivity* of the central dogma—that is, the assumption that the latter is uniquely sufficient to explain *all* aspects of the evolutionary process.[3] Nothing in what follows should be in any way construed as a *denial* of the existence and importance of natural selection, nor indeed of the validity of the extension of selection theory in contemporary work within the neo-Darwinist paradigm.

This comment applies as much to sociobiology as to classical Darwinism. Many trenchant critiques of sociobiology have focussed upon either or both the *reductionism* of its advocates, and the ideological *use* to which its theories have been put (Ho and Saunders, 1984; Rose, 1982a, 1982b; Rose, Lewontin and Kamin, 1984). While I do not intend to recapitulate these arguments here, I should emphasize (in line with the authors cited above) that a refusal to accept a general biological (and specifically genetic) determin*ism* does not equate to a refusal to accept the existence of specific processes of genetic determination—a point which seems frequently to be lost on the more rigid of neo-Darwinism's supporters, as well as on some of its more crudely environmentalist critics.

It is in point of principle (and of fact, so far as a non-biologist can evaluate this) quite possible to see the epigenetic approach as *compatible* not only with Darwinian theory in general, but also with neo-Darwinism, if this is interpreted, not as strict adherence to the 'central dogma', but as the *ensemble* of current work in the evolutionary sciences in the tradition of Darwin. Indeed, the neurobiological work of Changeux (1985), cited later in this chapter, adopts an explicitly epigenetic perspective from out of an equally explicit adherence to neo-Darwinism. At one level, then, the contrast I draw between epigeneticism and neo-Darwinism might be seen as more rhetorical and expository than as reflecting a deep theoretical incompatibility. Nonetheless, the differences of emphasis, interpretation and implication between neo-Darwininian and epigeneticist ways of theorizing evolutionary processes justify their being viewed as alternative paradigms in the sense of Kuhn (1970).[4]

The relationship between the other 'modern synthesis' and epigeneticism is substantially clearer, amounting in effect to total incompatibility.[5] Though younger, in its modern-synthetic form, than neo-Darwinism, it has a longer ancestry. I propose to call it the *neo-rationalist* paradigm,[6] to indicate its origins in the central debate in the recent history of psychology, which has been between the nativist and rationalist camp, as represented for example by N. Chomsky and J. A. Fodor, and the environmentalist-associationists in the tradition of J. B. Watson and B. F. Skinner.

The neo-rationalist paradigm, which has been so closely bound up with the rise of an inter-disciplinary cognitive science as to be virtually identified with it, has been consolidated in recent years into the dominant account of human behaviour and mental processes.[7] This account, combining the methodology of experimental psychology with the nativist and mentalist philosophy of traditional rationalism, and enriched by insights yielded by computer science, stands in a similar contrastive relation to epigeneticism as does neo-Darwinism: that is, as the dominant view to which (I shall argue) epigenetic (socio)naturalism is the only serious alternative. Thus, my argument presupposes the paradigmatic oppositon shown in Fig. 4.1.

The distinction between 'rationalist' and 'naturalist' psychologies was first employed in recent times by Fodor (1980), and his use of it appears to have commanded broad consent from both camps. The main bone of contention seems to be that 'rationalists' and 'naturalists' alike deny kinship with 'behaviourism'. In fact I shall suggest that behaviourist learning theory is best seen as part of the

Evolutionary biology	Cognitive science
Neo-Darwinism	Neo-rationalism
vs	vs
Epigenetic naturalism	

Figure 4.1: Alternative paradigms in evolutionary and cognitive science

'Classical synthesis' in early twentieth-century psychobiology, combining elements of both empiricism and naturalism within a fusion of reflexology and Darwinism. A consequence of the demise of empiricism as an adequate theory of human mental processes has been the break-up of the Classical synthesis, its replacement in psychology by the new synthesis of neo-rationalism, and a widening gulf between psychological and biological frames of explanation.

The term 'neo-rationalism' highlights the historical filiation of modern 'rational psychology' (Fodor's term) with the dominant tendency of Western thought, which has oscillated between 'empiricist' and 'rationalist' solutions to the same sort of questions, approached within the same presuppositional framework (see Chapter 1). As Fodor (1980: 64) himself notes, 'there's a long tradition, including both Rationalists and Empiricists, which takes it as axiomatic that one's experiences (and, *a fortiori*, one's beliefs) might have been just as they are even if the world had been quite different from the way that it is.' This 'contingency assumption' lies at the heart of both empiricist and rationalist theories, and is, I shall argue, enshrined in the modern synthesis in the guise of the competence–performance distinction.

If the intellectual forefather of neo-rationalism was Descartes, and the problematic which inspires it is shared with such figures as Plato, Leibniz, Hume and Locke, precursors of epigenetic naturalism can be found in the ideas of (among others) Heraclitus, Spinoza, Goethe, Hegel and Marx. The modern version of the epigenetic approach has been fashioned largely by such figures as J. M. Baldwin, G. Bateson and C. H. Waddington. The most systematic and prolific exponent, however, of (epi)genetic epistemology as a unified science of life and mind has been Jean Piaget—though certain aspects of Piaget's theory indicate that he, like his greatest mentor, Kant, uncomfortably straddles the rationalist *versus* naturalist divide. Be that as it may, Piaget's theory represents a fundamental reference point for the articulation of an alternative, epigenetic psychobiological paradigm.

Neo-rationalism: from reflexology to representation

The nature of cognitive science's 'modern synthesis' can best be introduced in terms of its research programme, which is directed at the attainment of a formally complete understanding of the nature of human mental processes, and (ideally) of the neural mechanisms underlying the translation of these mental processes into behaviour. Thus, cognitive science has decisively liberated itself from the behaviourist strait-jacket, which precluded any explanation which moved beyond the description of observable behaviour. Against Classical reflexology, which proposed a direct causal link between brain/nervous system and behaviour, the modern synthesis asserts the necessity of the study of mind as an autonomous level, possessing causal efficiency with respect to behaviour, and circumstantially constrained to, rather than epiphenomenal of, 'lower-level' neural and biochemical processes.

The modern synthesis, then, is anti-reductionist and mentalistic. It studies human behaviour in order to reveal the structures and processes permitting the operations of reasoning and the manipulation of symbolic information: its encoding, storage, retrieval, transformation and transmission. The fundamental premise of cognitive science is that human behaviour is rule governed and generative. That is to say, algorithmic rules intervene between different stages in coding processes in order to permit goal-directed problem solving.

Representation and computation

Representation is thus simultaneously the central problem, and, through its formal specification, the heuristic goal of neo-rationalist cognitive science. Information may formally be represented in different ways, in order to effect various procedures directed to specified goals. The mode of specification of the goals of procedures is itself representational: an end-state in a chain of transformations which may lawfully be derived from the syntactic properties of the formal rules constituting the representational system. This representational system is itself assumed to be represented, or instantiated, in the functional architecture of the brain and nervous system, and in neurochemical processes. Thus, the brain itself is seen as a representation of mind. This inverts the traditional reductionist view, in which mind was seen as a mere 'trace' of neural process. In the modern synthesis the brain is rather an embodiment or representation of mental process.

Further, it is no longer assumed that the particular instantation of mental processes embodied in human brains is either necessary, or constitutive of mind. Mental processes may equally be instantiated in artificial physical systems, such as computers. I shall refer to this assumption of the autonomy of cognitive process from neural process as the *contingency assumption*.[8] The contingency assumption underlies the distinction between simulations, on the one hand, and computational theories proper, on the other.[9] Simulations are, in principle, constrained to empirical evidence in the psychological or neurological domain which they are intended to 'model' in a more or less literal sense. Let us refer to simulation as Modelling$_1$. It can then be contrasted with Modelling$_2$, in which the 'model' is constrained, not to (neuro)psychological evidence, but to (internal) consistency criteria, and (external) correspondence to a formal object, such as a grammar.[10]

Although both Modelling$_1$ and Modelling$_2$ may be implemented in computer programs, the difference between them is (in principle) that the latter, rather than using computation as a means to the *description* of behaviour and of cognitive processes, sees computation as itself a fundamental *requirement* for cognitive theories.[11] The extension of the concept of representation to include the brain-as-a-representation is closely linked, therefore, to a reduction of representation to computation. The computational specification of mental representation presupposed by most work in cognitive science entails the subordination of representational content to computational form. The representational system is seen as being constituted by rules, of a formal and computational nature, and the elements upon which the rules operate are symbolic values, or strings of symbolic values, which are themselves defined over the formal rule-system.

Modelling$_2$, though it is bound by formal and theoretical considerations with respect to cognitive processes, is not bound to behaviour in the way in which Modelling$_1$ is, since it is assumed that (contingent) processing limitations, inherent in human neurobiology, intervene between mental processing and behavioural outcome.[12] Thus the contingency assumption also underpins the distinction between competence (constraint to a formal object guaranteeing internal consistency), representing the 'ideal' speaker-hearer, or more generally cognitive subject; and performance (constraint to 'contingent' limiting conditions on processing), representing the 'actual' human subject.

Grammar and psychology

It is evident from the foregoing that modern cognitive science is deeply indebted to theoretical linguistics. Indeed if one were to assign a birthdate to cognitive science, as good a one as any[13] would be the publication of Chomsky's (1959) review of Skinner's *Verbal Behaviour*. Chomsky makes a number of points, in this article and elsewhere, of which the most important for our current concerns are the following:

(1) Utterances in a natural language are fundamentally occasion-independent and non-stimulus bound, from a grammatical viewpoint. Syntax is not a probabilistic phenomenon, either in terms of linguistic context, or in terms of non-linguistic context.

(2) For any natural language, there is an infinite number of possible utterances (in a non-trivial sense; that is, one that is independent of particular lexical factors such as the infinite series of integers or proper names).

(3) This generativity of natural languages is a consequence of their grammars, and thus grammar is definitive of natural language.

(4) There are certain universal properties of the grammars of natural languages.

(5) These universals are not manifest in the actual forms of utterances/sentences (surface structure), but reside either in the grammatical deep structure or in the generative and transformational rules relating deep and surface structures. Thus neither

grammars nor the universal properties of grammars are phenomenologically accessible to subjects.

(6) To know a language is (definitively) to know its grammar (see 3).

(7) Neither grammar nor the universal properties of grammars are phenomenologically accessible to children acquiring language (see 5).

(8) Therefore, humans must be innately equipped with a pre-programmed Language Acquisition Device (LAD), specifying the universal constraints on grammars, and permitting the child to transduce the grammar of his/her native language from the 'degenerate' input of the actual utterances of other speakers.

This argument, to which I hope I have done sufficient justice, may be referred to as the prototypical 'rationalist deduction'. It both demonstrates the necessity for, and demarcates the research programme for, a new cognitive science of the mind, its laws and properties: a universalistic, nomological and individualistic science of mind.

Modularity

The modern synthesis, resting upon the twin pillars of the contingency assumption and the rationalist deduction, is perhaps best typified in terms of its answers to two perennial questions. The first question is: what is the relationship between mind and brain? The second question is: what is the relationship between knowledge and experience? To these two questions, the modern synthesis answers, respectively, modularity and maturation.

The computational theories which have the widest currency within the framework of the modern synthesis are (ideally) constrained to well-formulated theories of particular domains. The paradigm for such theories is provided by natural language grammars. The logical next step is to search for quasi-grammatical, generative formalizations of non-linguistic domains. Thus the programmatic evolution of cognitive science has been dominated by the search for self-consistent, autonomous and generative 'grammars', of action, of vision and so forth. Within this paradigm, goal and method coincide in the imperative of formalization: since computability is presupposed by the paradigm, only theories which constitute formal computational descriptions of the domain under study are acceptable candidates. The paradigm therefore introduces closure, at two distinct levels.

Closure$_1$. Only formally coherent (thus closed) systemic descriptions are deemed scientific, and mechanisms defined over formally unspecified variables are unacceptable. This condition upon theory-building violates a fundamental 'open-ness' presupposition of science, that models should be non-prescriptive of mechanisms that cannot be evidentially supported. The common resort within the paradigm is to default characterization of values yielded by (actually) unknown mechanisms. This is a perfectly acceptable procedure, but it is in principle incompatible with the claim that model and reality coincide. In effect, the neo-rationalist paradigm is trapped within an irresolvable oscillation between correspondence and coherence models of truth.

Closure$_2$. The condition that models should be self-consistent (coherent) *within* domains, together with the aim that such models should be exhaustive (fully explanatory), not only inevitably leads to *ad-hoc* characterization of unknown variables/values, but also leads to non-isomorphism of terms *between* domains. The net effect of Closure$_1$ + Closure$_2$ is to induce a preferred meta-theory in which not only cognition as a whole, but also particular cognitive processes within specific domains, are self-enclosed.

The result is a modular theory of mind, which, when given a physicalist interpretation according to the principle of the brain-as-a-representation, assumes the form of a modular theory of brain function. Such a perspective is articulated by, amongst others, Chomsky (1980), who advances a theory of innate faculties of mind, to each of which correspond different 'mental organs'. Marshall (1980), in his response to Chomsky's proposal, has noted the similarity between this and the eighteenth/nineteenth-century 'organology' of Franz-Joseph Gall. Fodor adopts and qualifies this comparison in his book *The Modularity of Mind* (1983), in which he develops an account based upon modular, domain-specific, perceptual input analysing devices.[14]

It is important at this stage to note that: (1) Fodor's suggestions regarding modularity differ substantially from those of Chomsky, as Fodor himself notes; and that (2) they do not necessarily stand or fall together with his general positions regarding either the Representational Theory of Mind (RTM) or the Language of Thought (LOT) (see Chapter 2).[15] The reason for this is that

Fodor's arguments regarding input analysers are couched not in (Chomskian) terms of innate propositional contents, but in terms of neuropsychological mechanisms based upon automatic (or automatized) parsing principles.

As Schilcher and Tennant (1984, Chapter 4) argue, however, the construction of parsers delivering structural descriptions of sentences in (formally) decidable languages does *not* require the same kind of assumptions regarding 'knowledge' (tacit or not) that are necessitated by the Chomskian account of the construction (or evaluation) of generative grammars. Schilcher and Tennant present this argument in terms of computational power. The argument is that Chomskian grammars conspicuously fail to constrain the class of possible languages to those which may be parsed by sub-recursive devices. Since a Universal Grammar (UG) instantiated in LAD will need *otherwise* to constrain its outputs (actual grammars), such constraints are built into LAD as a computationally powerful set of non-inducible rules or principles ('knowledge of language').[16] If, however, the class of possible languages is restricted, not by UG but, in virtue of the desirability of computational non-complexity, to context-sensitive and/or context-free languages, then such knowledge is unnecessary.[17]

The same argument may reasonably be assumed to hold for other stimulus types (or 'informational natural kinds')—that is, that their *recognitory processing* is pre-epistemic in comparison to whatever computations might putatively be necessary to stimulate their generative structure.[18] This is not an argument applying solely to reception. Similar arguments apply in principle to production. Actual speech production, as Langacker (1987) argues, need not be thought of in terms of the deployment of 'knowledge of a grammar' in the usual sense of a formal or constructive device. Although the production of appropriate and grammatical utterances does imply various types of knowledge and of cognitive activity, their objects are not, Langacker suggests, the kind of strictly linguistic units assumed in generative grammars.[19]

Although Fodorian input analysers are conceived in terms of computation, the pre-propositional nature of the contents to which they are dedicated distinguish them from the mental representations which are the more usual objects of Fodor's (and other) neo-rationalist theories. Because input analysers are not defined over (computationally and semantically interpreted) representations, but instead deliver recognitory-structural information to 'central' processes, it is not strictly necessary to assume that the central processes are themselves computational (in the $Modelling_2$ sense of the word) in nature.[20] Thus, even if, as Fodor has argued (see below), one cannot have LAD without RTM and LOT, one can nevertheless accept input analyser modularity without accepting any of LAD, LOT or RTM. In other words, Fodor's 'encapsulated input analysers', whether or not conceived computationally, are quite compatible, in principle, with a non-nativist and non-computational (in the terms, that is, of neo-rationalism) theory of cognition (in the sense of conceptual knowledge and belief, or what Fodor calls central processes).[21] The degree of innate specification of the input analyzers themselves would, on such an account, be an empirical question. Without anticipating later arguments in full, one might further suggest that 'encapsulation' and 'automatization' are better seen in terms of developmental processes (in both ontogeny and phylogeny—see Chapter 3) than in terms of 'rationally designed' interfaces.

There is, then, no logical compulsion to dismiss the thesis of modularity, if this is understood in Fodor's (1983) terms, even if one does not accept the RTM/LOT theory. Indeed, the modularity thesis, implying a vertical division of neuropsychological (perceptuo-motor) function, is no less plausible in evolutionary terms than theories suggesting a horizontally specified cerebral functional architecture (e.g. MacLean, 1972). Modularity certainly does not imply strict independence (as opposed to relative autonomy) of different input (or output) systems. Rather, mind-brain (at least, that part of it which constitutes the organism-world interface) is conceived in terms of co-operatively interacting functions, each specialized to different perceptuo-motor domains, and each (perhaps, though not necessarily) pre-programmed with its own developmental timetable and processing characteristics.

In what follows I shall be concerned not so much with Fodor's 'input analyser' version of modularity, which I have argued to be theoretically neutral with respect to the neo-rationalist programme, as with the Chomskian conception of modularity, in which something like knowledge or belief[22] necessarily figures as a property of whatever is hardwired in the modules. Such a conception of modularity, unlike the 'input analyser' version, raises in a very direct way the issue of how such modules interface with the 'central' and 'explicit' processes involving concepts and propositional attitudes. Put more formally, this resolves into the question: is there a LOT, computationally defined over the outputs of cognitive modules, whose specification is the goal of RTM? In Chapter 2, I advanced some arguments against such a position; in what follows I shall provide some further criticisms, based upon an examination of its consequences, and attempt to formulate an alternative.

Growth and learning

The new organology of neo-rationalism is also a maturationist theory of development, in which there is little room for learning in the traditional sense. Chomsky (1980: 14) goes so far as to state that, at least as far as *knowledge* is concerned, 'it is rather doubtful in fact, that there is much in the natural world that falls under learning'. He relegates learning theory, as traditionally understood in psychology, to 'the study of tasks and problems for which we have no special design, which are at the borders of cognitive capacity or beyond, and are thus guaranteed to tell us very little about the nature of the organism'.

I shall return to the problem of evolutionary design which this theory poses. To continue, however, the appropriate model, within neo-rationalist theory, for cognitive development is growth, rather than learning. This maturationist position appears, at first sight, to fall in the mainstream tradition of American developmental psychology, as represented in the early twentieth century by such figures as Gesell and G. Stanley Hall; a tradition which repudiated the socially oriented epigeneticism of Baldwin. Unlike traditional maturationism, however, the new synthesis is anti-naturalistic and anti-behaviourist. Its method, as I have shown, is formalist, relying upon computational models, and its focus is upon mental objects rather than on behaviour.

Traditional maturationism, by contrast, was both naturalistic in method (relying upon observation) and easily accommodated to a behaviourist paradigm. The possibility for such an accommodation was given by the universal acceptance, in the first part of this century, of the reflex arc as simultaneously the fundamental neuropsychological unit (the unit of association); the paradigm case of organism-environment transaction (the S-R link); and the basic mechanism of learning (through either Pavlovian or operant

(a) The classical synthesis: reflexology
(b) The modern synthesis: cognitivism

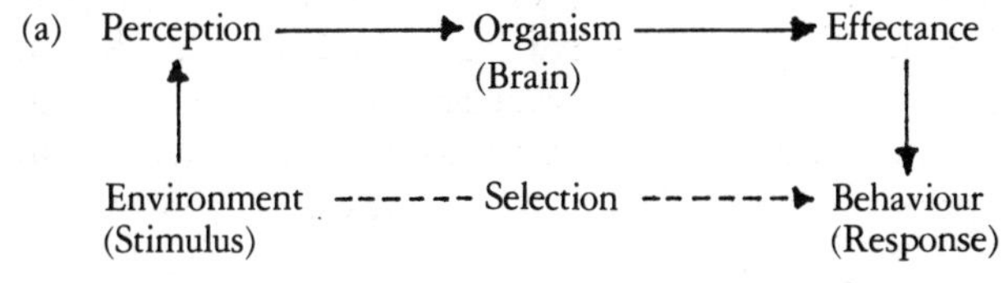

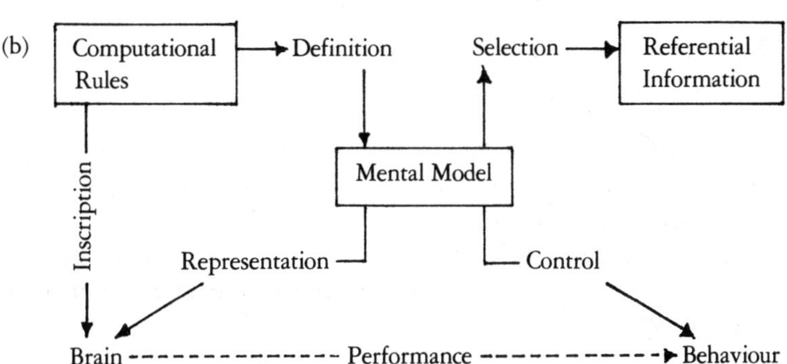

Figure 4.2. From reflexology to representation

conditioning). Such a paradigm, essentially empiricist in nature, can combine behaviourist learning theory with the recognition that certain neural circuits are maturationally established, rather than established strictly through experience. It contrasts with the modern synthesis insofar as the latter sees Classical learning as peripheral, is not founded upon the reflex arc, and substitutes internal constraints for external ones. Figure 4.2 illustrates the difference between the old and the new paradigms.

Figure 4.2(a), representing the Classical—associationist and reflexological—synthesis, is simply a slightly elaborated version of the sort of 'Stimulus-Organism-Response' diagram familiar from traditional accounts of learning theory. The important point to note is that, according to the view that it represents, it is the *environment* which selects (or 'elicits') behaviour, through either association-by-contiguity (Pavlovian conditioning) or reinforcement and extinction (operant conditioning). Developmental change can occur either through the maturation of pre-established neural circuits, or by means of the establishment of new circuits through learning.

Classical learning theory, in fact, provides a perfect correlate, in terms of change within the lifespan of the individual organism, to the Darwinian theory of natural selection of intra-specific variations in a phylogenetic timespan. Both theories account for increased adaption on the part of the organism by ascribing a primary role to the environment. The theories are also complementary insofar as maturational developmental change, whether morphological or behavioural, can be accounted for by means of natural selection—as in the case of 'instincts'—leaving learning theory to account for the residue of ontogenetic change. Further, this learned residue is assumed to be of an extremely plastic nature—it is reversible and completely inconsequential as far as the genes transmitted by the individual are concerned.

Classical learning theory is therefore predicated upon a division between instinctual and learned behaviour which precisely parallels Weismann's barrier in neo-Darwinian theory. Finally—and this is important to note—both neo-Darwinism and reflexology assign a central place to behaviour. Though much of the subject-matter and evidence for Darwinian natural selection concerns morphological characteristics, it should not be forgotten that the mechanism of selection operates upon functioning organisms, upon the behavioural correlates or consequences of structural features, and not upon the structure itself. It is precisely this which distinguishes Darwinian from Lamarckian evolutionary theory.

By contrast, the cognitivist theory represented in Fig. 4.2(b) locates the control of behaviour in a 'mental model', defined in terms of the operation of computational rules.[23] The representation, in the mental model, of a particular, local and current 'world-state', involves the *selection* of relevant *referential information* from environmental input. The selection process, however, is also a *construction* process, insofar as the value of referential information is itself defined over the computational rules: in this sense, the only *actual* 'environment' is the mental model. The real environment enters into the mental model (and into the theory in general) only as selected referential information (values or parameters upon which the computational rules operate).[24]

An equivalent process is assumed to operate in the brain. Where the mental model is defined in terms of the computational rules, the brain is assumed to have these rules pre-programmed or inscribed in it. The access which the organism has to the environment is solely in terms of the mental model 'represented', at any given moment, in the brain. Further, given the contingency assumption, the systemic unit labelled 'brain' might more generally be labelled 'hardware'—denoting any physical system capable of performing the necessary computational work. *Which* particular physical system is chosen to implement the model is relevant only insofar as the performance characteristics of the system will affect the actual behavioural output.

Radical nativism

The computational rules are therefore in no sense derived from the environment; rather, they are innate formalisms enabling the system to select relevant features of input in order to construct the environment. The selection and construction of the environment by the system, as opposed to the selection of behaviour by the environment, most fundamentally distinguishes the new from the old synthesis. This selective constructivism applies both to ontogenesis and to mature functioning, in that innate formalisms enable the subject to select from environmental input only those features which are either developmentally or currently relevant to the construction process in hand.

Such a conception raises the difficult problem of how it is that the cognitive system comes in the first place to 'know' which features of the 'real' environment are those which are necessary and sufficient for the construction of the representational environment. This question is as much philosophical as psychological—indeed, more so, for the psychological and computational solution is implied by the theory: such 'knowledge' is simply built in. Such a pragmatic solution, however, is unable to escape the charge that, although it explains everything, it thereby also explains nothing.

A more rigorous philosophical justification for such a procedure has been offered by Fodor (1976, 1980), who exends the 'rationalist deduction' a further, and crucial, step. Computational models of mind of necessity require that there exist a specifiable set of (interpretable) symbolic values. These values are defined over the rule system, which is itself non-derivable from empirically observable reality. The operations of the rules yield representations, and these representations, according to RTM, *are* mental states. Fodor further argues that there can, in principle, be no adequate means of specifying the 'objects' of mental states in terms of the languages of the special sciences such as physics: thus, it must be assumed that the system-internal values, as well as the formal rule systems, are innate; either as atomic 'primitives', or in the form of further rules such as 'meaning postulates' (Carnap, 1956). The computational approach appears therefore to be committed to a radical nativism, predicated upon a dualistic distinction between mental objects ('lexemes' and expressions in LOT) and the actual objects of the real world.[25] This neo-rationalist argument is used by Fodor to defend a general philosophical approach to psychology which he calls 'methodological solipsism', consisting essentially in the proposition that the study of cognitive processes is to be seen as completely autonomous from the study either of the cognized environment or of developmental history.

The 'methodological solipsist' stance also highlights another basic feature of the neo-rationalist position: that is, its tendency to cast the problem of knowledge in terms of the mental states of individual subjects, rather than, or in opposition to, knowledge being conceived as a social process and product. To this extent, Fodor's methodological solipsism is simply an extreme variant of a wider tenet of modern cognitivism, which may be referred to as 'methodological individualism'.

As well as a philosophical and psychological aspect, however, the neo-rationalist synthesis also possesses a biological aspect, which can be stated in terms of the problem of how such a representational system might have emerged from biological evolution. It will be recalled that the 'Classical synthesis' provided a theory of learning which was, in its determinate features, theoretically interlocked with Darwin's theory of natural selection. No such theoretical concordance with evolutionary theory can be claimed for neo-rationalism, for two fundamental reasons. First, because of the cognitivist assumption that the environment is selected by the subject, rather than vice versa. Second, because the competence–performance distinction, rooted in the contingency assumption, accords to behaviour a secondary role *vis-à-vis* the computational mechanisms governing and controlling it.

Further, the argument-to-radical-nativism provided by Fodor

has implications for evolutionary theory every bit as negative as those for cognitive developmental theory. In the neo-rationalist synthesis, 'representation', or computational knowlege, is seen as distinct from knowledge of the physical properties of the universe, which are considered as phenomenologically inaccessible. This conception of representation, which depends upon a strict distinction between mind and nature, is indicative of the Cartesian dualist inheritance of modern neo-rationalism.[26] Despite, or because of, the mentalistic, rather than empiricist, orientation of the computational approach, it remains unable to account for the emergence of mind from nature, and is therefore in many respects profoundly anti-evolutionary. It should come as no surprise, then, that Chomsky, for example, repeatedly stresses the (innate) discontinuity between human mental processes and those of other species, and the biological uniqueness of manifestations of human mentality such as language.

It is my argument, then, that although neo-rationalism does not explicitly disavow evolutionary theory, its deep structure provides no compelling reason for accepting it, and indeed may be read more as an argument in favour of special design. Thus, although its nativistic and maturationist biases appear at first sight to afford an easy accommodation with biology, the fact is that no mechanism is provided to account for the origins of representation in naturally evolving life. In this sense, neo-rationalism merely reflects the predominant, but implicit, *biologism* of contemporary psychology, rather than representing a genuine opening to psychobiological theorization.

'Ecological' alternatives

The appeal of cognitivism has never been universal, and even within cognitive psychology the computational approach has been subject to criticism on a number of counts. These range from its lack of concern with context and with actual behaviour, to its 'dehumanizing' equation of machine with human intelligence, and the difficulty of deriving intentionality from computation.[27] In this section, I shall outline alternative approaches which draw upon (mainly) naturalistic evidence and methodologies to construct accounts of human and non-human behaviour in their 'ecologically valid' contexts. I shall also attempt to evaluate the extent to which such approaches meet the implicit challenge of the neo-rationalist synthesis: to develop a convincing alternative account of representation and its development.

Ecological validity (1): context, rules and meanings

As has been noted, a recurrent feature of the computational approach is its lack of concern with behaviour as this actually occurs in natural situations. Cognitive science does, indeed, sometimes employ arguments based upon laboratory experiments with human subjects. Frequently, however, the 'data' are pre-theorized in a formal sense (in the way in which a grammar might count as data). This lack of concern with behavioural data opens the computational approach to the fundamental objection that, whatever else it is doing, it simply is not producing explanations of how subjects actually perform tasks involving reasoning, inference and so on in natural settings.

This criticism of lack of ecological validity has been levelled at the entire tradition of laboratory-based experimental psychology (Cole *et al.*, 1971; Neisser, 1976), but it has greatest force in relation to computational theories. It is significant, furthermore, that the ecological validity argument, when applied to human psychology and behaviour, has been closely associated with attempts to reinstate the social dimension of cognition, particularly in relation to its development. There has in recent years been a growing recognition by developmental psychologists of the importance of studying the social-ecological matrix within which human development naturally occurs (Bronfenbrenner, 1979; Butterworth and Light, 1982), and a renewed emphasis upon the socio-communicative origin and elaboration of cognitive processes (Bruner, 1975; Camaioni, de Lemos *et al.*, 1985; Donaldson, 1978; Hickmann, 1987). These developments have led, too, to an appreciation of the nature of the experimental setting itself as a 'socio-dialogic context' (Karmiloff-Smith, 1979; Walkerdine and Sinha, 1978). These recent debates in developmental psychology, over the relative weight to accord to intra-individual and inter-individual cognitive processes, echo the Marxist-derived critique Vygotsky offered, half a century ago, of Piaget's genetic epistemology.

The basic issue at stake here concerns the extent to which it is either possible or useful to attempt a theorization of individual mental representations and processes, without reference to the interpretation and evaluation of social actions by sign-using communities. This problem is, of course, fundamental to the human sciences, and strikes at the heart of the claims of the computational approach to scientificity. Contemporary approaches to the psychosocial and philosophical analysis of action (Apostel *et al.*, 1987; Cranach and Harré, 1982; Miller, 1987) increasingly stress the dialectical interdependence between actors' mental states (beliefs, wants, intentions etc.) and interactants' evaluations of act-performance and outcome. The fundamentally communicative epistemology implied here, rooted both in critical social theory (Apel, 1980; Habermas, 1971), and in pragma-semiotic and text-linguistic theories (de Beaugrande and Dressler, 1981; Castelfranchi and Poggi, 1987; Eco, 1984; Parret, 1983, 1985; Petöfi, 1987), is distinctly orthogonal to the main stream of cognitive theory in general, and not only in its dominant neo-rationalist and computational manifestations. This inevitably raises the question of whether cognitivist and socio-pragmatic approaches represent—as Marková (1982) implies—incompatible, or—as Nuyts (1987) argues—potentially complementary, perspectives (see also de Gelder, 1985).

Because this entire book can be seen, in a sense, as a meditation upon this question, it would be artificial to force a definitive answer here. Two comments, however, seem in order. First, it is apparent that the methodological individualism inherent in the dominant, neo-rationalist paradigm is by no means the only possible broad perspective for cognitive and communicative theory. This by no means implies that a single, coherent and widely agreed social action-theoretic paradigm currently exists. However, advocates of widely varying positions, ranging from naturalistic (Bhaskar, 1979) to hermeneutic-constructivist (Rommetveit and Blakar, 1979; Shotter and Newson, 1982); and from Wittgensteinian (Hamlyn, 1978, 1982) to dialectical materialist (Sinha, 1982a, b; Wertsch, 1985), all find some measure of common ground in their insistence upon the importance of intersubjective agreement as a key attribute of human rational thought in the widest sense.

Second, the exclusive focus on formalization of most current cognitive science is unconducive, in practice even if not in principle, to systematic investigation of the pragmatic negotiation of meaning (signification) within episodes of social exchange. It is a moot point whether rigorous formalization of pragmatic 'rules' is in principle possible. But so long as cognitive science relegates such phenomena—most, that is, of what is constitutive of human social existence—to a separate category of 'performance', it will be unable to effect a fruitful rapprochement with many of the most exciting developments in the contemporary study of language and communication in its social context.[28]

Ecological validity (2): direct perception

The ecological validity argument is not restricted to the issue of the interpretation of human behaviour within its social and communicative context. In fact, for most psychologists the phrase is predominantly associated with an approach which derives in a more straightforward fashion from the study of the adaptive circuits linking organisms and environments within ecosystems. The psychobiological premises of what is generally known as the 'ecological approach' are stated by Shaw and Turvey (1980: 96) in the following terms:

> logically dependent ... animal-environment synergy (or reciprocity) ... identifies (a species of) animal and its environment as reciprocal components comprising an (epistemic) subsystem.

I shall attempt to expand this somewhat opaque statement in a discussion below of the concept of behaviour. For now, the salient features of the 'ecological approach' upon which I shall concentrate are its denial of the necessity of invoking any kind of concept of representation in the explanation of behaviour, and its employment of the alternative theory of direct perception developed by J. J. Gibson (1904–1980). In this brief exposition, I shall concentrate upon Gibson's later formulations of his theory (Gibson, 1979).[29]

Gibson's life-long research interest was in the perception of visual space, as a structured, textured and complex psychobiological unity. His method of approach led him to a radical reformulation of Classical psychophysics, in which the concept of 'ambient information' replaces previous terms such as 'stimulus' and 'sensation' which were associated with the reflexological model; and to a decisive break with cognitivist models of perception in which the 'percept' is constructed by the subject from degenerate or inadequate information.

Gibson held that the visual field is directly specified by the structure of light, the flow of optical texture. Classical psychophysics, he argued, based upon Newtonian optics, upon atomistic notions of the 'stimulus', and upon untenable philosophical distinctions between 'primary' and 'secondary' qualities, was unable to reconcile the constancy of our perception of objects with the apparent ambiguity, in real contexts, of the isolated 'cues' which indicate or 'represent' the objects. Hence the cognitivist argument in favour of an inferential supplementation of psychophysical stimulation. Gibson's alternative solution to this problem was to suggest that a different psychophysics, a non-Newtonian or *ecological optics*, was required, which would obviate the need for cognitive supplementation of perception.[30]

Ecological optics departs from the geometrical and atomistic abstractions of Classical optics by invoking the priority of *background* in specifying 'objectness' within a textured (rather than empty) space. Ecological space is conceived as a surface, textured by variate patterns of information, whose objective structure is specified by variables within the ambient light (gradients of shade, scatter, intensity etc.) The structure of light therefore contains all the information necessary for the 'pick-up' of invariant properties, such as those involved in motion perspective and object constancy, without the subject having to do any further representational or mental-inferential work.

The theory of direct perception also views the senses as exploratory, attentional organs, of an organism which is fully, actively engaged with its environment, by means of locomotion and through the variable articulation of body parts and orientation of sense organs. Thus the organism is itself a part of the niche. In this sense, the theory of direct perception depends upon the wider propositions of the ecological approach: that a niche is a negotiated, ordered, spatio-temporally structured relationship between organism and habitat, in which behaviours are in part transformative of the environment to which they are adapted.

In his final work (Gibson, 1979), Gibson introduced a further, and extremely important, concept into the theory of direct perception: this is the notion of *affordance*. As the name implies, 'affordances' consist in the properties of objects by virtue of which these objects serve to sustain or permit the actions of the organism. The concept of affordance, then, fixes the theory of direct perception within the wider propositions of the ecological approach. Affordances, maintained Gibson, are themselves directly specified by ambient optical information, actively 'picked up' by an organism which is, in every sense, *interested* in its environment. Objects exist, for the organism, as meaningful parts of the phenomenal world, precisely because they afford action—they are edible, climb-able, graspable etc.—and these affordances are 'grasped' by the perceptual system of the organism in its direct engagement with the optical array. As Costall (1981) notes, the theory of affordances is an important new departure in Gibson's work, for it points toward a functionalist theory of perception in relation to adaptive action. As I shall argue, however, the theory of affordances is also a Trojan horse within the Gibsonian epistemology.

To what extent, then, does the ecological approach offer a solution to the problem of representation? First, it escapes the central problem of rationalist (and empiricist) theories, that of the relationship between mental or internal objects, and real objects, by denying the existence of 'percepts' as 'copies' of real objects. Objects are directly perceived, and the information picked up in direct perception is neither a mediating construct, nor an isomorphic 'picture' of the world, but a *real structure* which specifies objects by virtue of its internal, higher-order variables: 'the question is not how much [the retinal image] resembles the visual world, but whether it contains enough variation to account for all the features of the visual world' (Gibson, 1956: 62).

These variables are defined, not in terms of the special predicates of theoretical physics, but in terms of local, 'human scale' (or organism scale) psychophysics (biophysics). To this extent, the ecological approach offers both a real alternative to neo-rationalism, and a formidable challenge to computational approaches in particular, and cognitive science in general.[31] Nevertheless, the ecological approach cannot, by definition, offer a solution to the problem of what have traditionally been called 'higher cognitive processes'—those processes operating, not upon the proximal environment, but upon symbolic representations. Even granted 'direct perception' of the proximal environment, can we then simply graft a computational theory of mind onto it? This seems obviously unsatisfactory, for it leaves an unbridgeable gap between the world we live in, experientally, and the world we think about, remember, make plans and draw inferences about. Of course, for the 'methodological solipsist' this presents no problem—but this takes us back, in an endless loop, to the beginning of the argument.[32]

A radical Gibsonian would, presumably, hold that representation is an unnecessary concept *tout court*, adopting a neo-behaviourist position and ruling 'mind' out of the bounds of science. As yet however, behaviourism has singularly failed to provide an adequate riposte to Chomsky's critique, and such a step would undoubtedly be regressive. A more interesting step would be to build upon the notion of 'affordance', suggesting that conceptual representations consist of mnemonically stored 'abstractions' of affordances. In Chapters 3 and 5, I suggest something similar, but as I shall show, this has fatally serious consequences for direct perception theory, at least as applied to adult human subjects.

Much of the human environment consists of artefacts, which indeed afford actions, but many of which resemble natural kind objects not even remotely. In what sense is it plausible to say that the functional affordances of a car, for example, or a telephone, are 'directly specified' in the optical array? A car-seat, or a steering-wheel, might be said to 'afford' appropriate actions, in a Gibsonian sense, but these sort of affordances, cannot, however you add them up, yield knowledge of how to drive a car, or what a car is: that is, of the socio-cultural 'affordances' of cars as complex artefacts within a social system. Thus, 'direct perception' cannot specify *knowledge* (or belief), whatever else it might specify (e.g. segregated objects of perception and attention). Even more seriously, this argument demonstrates the questionable status of 'affordance' as a purely *perceptual* category.

In fact, the argument can be applied to natural kind as well as to artefactual objects. To the untutored eye, a piece of rock might 'afford' merely the actions of kicking, using it as a missile, or an interesting paperweight, etc. To a geologist, however, it might 'afford' a hypothesis, evidence for a theory, an indication of a valuable resource etc. In effect, the whole notion of 'affordance' necessarily involves, at least for adult human subjects, propositional attitudes, significations and representations. Its introduction into ecological psychology means that the theory is no longer restricted to perception, but touches upon cognition too.

Unfortunately, ecological psychology has no theory of representation. Thus, if the concept of affordance is to serve as a starting point for such a theory, that concept will certainly require much critical revision. If neo-rationalism fails to provide a link between representation, evolution and behaviour, so, ultimately, does the theory of direct perception.

Behaviour, representation and evolution

There are a number of possible ways in which behaviour may be described. One way might be to limit one's descriptive terms to those which refer solely to the organism, or to its body parts, or to their movements relative to one another. Such a description, however, seems inherently unsatisfactory as a description of behaviour, although it is a legitimate enterprise in itself. It seems more appropriate, in such a case, to speak of the physiology of movement, and while this may be an indispensable complement to the description of behaviour, the two are essentially different. For behaviour is motivated and meaningful, consisting not merely in movements, but in movements in relation to an environment. The movements of even the simplest of life forms are adapted to a particular niche. Even where such adaptations appear merely to consist in a one-to-one causal link between a specific type of environmental event and a specific type of movement, as in the S-R links of learning theory, or the fixed action patterns studied by ethologists, a full *behavioural* description necessarily involves reference to the environmental event structure, as well as the movement structure.

As we move to the consideration of more complex behaviours, the connection between behavioural and physiological descriptions becomes increasingly remote. Flight and predatory pursuit both involve rapid locomotion, but the meaning and 'cause' of the movements is different in the two cases. Ethological studies are often concerned with the detailed specification of particular *combinations* of movement, posture and attentional orientation which constitute a particular behavioural routine; but the degrees of freedom in the co-articulation and sequential organization of micro-behaviours within macro-behavioural units increase rapidly with increased phylogenetic complexity. Indeed, it is the relative freedom of behaviour from 'stimulus', and of behavioural goal or end from the detailed motoric means of attaining that goal, which typifies 'intelligent adaptation'.

At a minimum, then, the description of behaviour necessarily involves reference to an environment to which that behaviour is adapted. This is, of course, the basic tenet of the ecological approach. However, the description of intelligent behaviour requires, additionally, and as we have seen, reference to the *representations* of the environment which the organism utilizes (at the highest level, intentional states such as beliefs and desires), which specify a complex repertoire of behaviours which the environment actually or potentially affords, and which contribute to the maintenance or expansion of adaptive fit between organism and environment.

An essential feature of behaviour is that it is *active*, going beyond the bounds of the organism to affect the environment and change it—even if this change simply consists in the substitution of one spatio-temporal segment of the environment for another, as in locomotion or migration. This active nature of behaviour was stressed by Piaget, who defined behaviour as 'all action directed by organisms towards the outside world in order to change conditions therein or to change their own situation in relation to these surroundings.' (Piaget, 1979: ix)

Piaget makes it clear that autonomic self-regulation, such as breathing, is not behaviour, although it is quite clearly adaptive. He goes on to state that 'behaviour is teleological action aimed at the utilization or transformation of the environment and the preservation or increase of the organism's capacity to affect this environment', and that, further, 'the ultimate aim of behaviour is nothing less than the expansion of the habitable—and, later, the knowable—environment' (Piaget, 1979: x, xviii.) It is in this sense that he saw behaviour as the *motor* and leading edge of evolution and development.

In what sense is behaviour adaptive?

The straightfoward neo-Darwinian answer to this question is: behaviour is adaptive insofar as it contributes to inclusive fitness. But there is another common sense in which we consider behaviour to be adaptive: that is, when it is *appropriate* to a particular situation or local environmental segment. For adaptation in this latter sense, Piaget (1979) frequently uses the expression *adequation.*

The two senses of adaptation are, to be sure, related: the inclusive fitness of an organism which consistently behaved inadequately would, indeed, be poor. But a particular behaviour confers selective advantage only in relation to a given context of occasion, for which its realization is appropriate (adequate) in terms of local goals. A fundamental deficiency of neo-Darwinism is its inability to conceptualize this distinction. This deficiency is related in turn to neo-Darwinism's difficulty in explaining the central paradox of evolution: why, if an organism is 'adapted' in the sense of inclusive fitness, should evolution occur at all? Where neo-Darwinism invokes purely exogenous selective 'pressure', an epigenetic approach sees the dynamism of evolution as consisting in an endogenously expanding circuit of adaptation, driven by the exploratory fallibility of behaviour. That is to say, it is the possibility of 'inadequacy' that underlies the leading role of behaviour in evolution and development.

Behaviour is goal-directed, and intelligent behaviour is representationally mediated. It is characteristic of representations that they can be both erroneous and impoverished. Behaviour, as Piaget emphasized, is defined foremost in terms of goals, and only secondarily in terms of degree of coincidence between the goal of the behaviour—its 'intended' outcome—and its actual outcome, which are frequently divergent; or between 'representation' and 'reality', which are also frequently divergent.

Not only can behaviours be both situationally inadequate and/or unsuccessful; but also behaviours, and overall behavioural repertoires and strategies, can result in environmental consequences which, though a part of the causal circuit linking organism to environment, are not the proximal goal of the behaviour. A 'path' may, for example, be an unintended consequence of repeated locomotion from one place to another, but it is nevertheless a useful one. Ecologists emphasize that species shape their niche, as well as being shaped by it, and such environmental shaping is partially constitutive of the species' survival strategy.

Further, such shaping, for all species, including our own, can introduce distal consequences—food shortage, erosion, pollution, competition with other species—which are outside the initial circuit of adaptation. That useful path may be disadvantageous if a predator cottons on to the idea of lying in wait at certain times of the day or night. These disadvantageous consequences will require a further set of adaptions if survival is to be assured and extinction averted—or, more generally, unpleasant difficulties surmounted.

Accommodation and assimilation (1): Baldwin

Such secondary adaptations were referred to by the epigenetic psychobiologist James Mark Baldwin[33] (1861–1934) as *accommodations*, and he saw them as playing a crucial role in both phylogenesis and ontogenesis. Accommodation constitutes a mechanism of *active selection* on the part of the organism, oriented to enhanced adequation, which prescribes a new, or wider, range of dynamic adaptation, rather than conferring a fixed selective advantage. The concept of accommodation sharply contrasts, therefore, with such notions as that of the Evolutionary Stable Strategy (ESS) favoured by neo-Darwinists: accommodations are active, *organismically* selected responses to induced environmental demands, whereas ESSs are *environmentally* selected from random behavioural variation.

The cumulative process of accommodation leads to what Baldwin called 'organic selection', which he contrasted with the negative and purely limiting role played by natural selection in evolution. Organic selection is a process independent of natural selection, by which individuals extend their adaptive range by accommodation, thus surviving to 'permit variations oriented in the same direction to develop through subsequent generations, while variations oriented in other directions, will disappear without becoming fixed' (Baldwin, 1897: 11). Thus, 'individual modifications or accommodations supplement, protect or screen organic characters and keep them alive until useful congenital variations arise and survive by natural selection' (Baldwin, 1902: 173). This effect is still referred to as the 'Baldwin effect'. Baldwin conceived of it as playing an increasingly decisive role in evolution, since he maintained that the range of accommodatory plasticity, and hence the importance of organic selection, increased with phylogenetic complexity. Thus, organic selection, combined with imitation (see below), increasingly direct, rather than follow, the course of natural selection. As Piaget (1979) noted, Baldwin's theory of organic selection anticipates the concept of 'genetic assimilation' (Waddington, 1975), and the mechanism of 'phenocopy', whereby a phenotypic adaptation is replaced by a variation in the genotype (Ho, 1984).

The process of accommodation, for Baldwin, also underlies the ontogenetic emergence of higher mental processes. Accommodation permits the integration of behaviour with a complexly structured and variably respondent environment, including the behavioural actions of other individuals in response to the actions of the infant. A key example given by Baldwin is the development of imitation, which he sees not as a passive 'copy' of the model, but as an active effort by the infant to overcome the resistance of her own body. The first accommodations are those which 'select' from among the infant's habitual actions in order to repeat pleasurable experiences.

These 'circular reactions' (see also Chapter 3) form the basis for subsequent accommodations to the actions which they provoke in others, giving rise to imitations. Imitations are subsequently stored as an associative 'net', enabling new events and objects to be *assimilated* to the products of past accommodations. According to Baldwin, the earliest representations are therefore mnemonically 'fixed' accommodations. These early representations, however, are inadequate to the full complexity of reality, which resists assimilation to the infant's primitive intentional accommodations. This leads to what Baldwin calls an 'embarrassment', the awareness of 'an inevitable mismatch between what the infant expects and wants and the behaviour of objects' (Russell, 1978: 54). The dialectical motor of development then, consisted for Baldwin in precisely what I have called 'the exploratory fallibility of behaviour'. Subsequent accommodations are also necessitated by the interventions of adults, who draw the child's attention to the inadequacy of her accommodatory representations and actions, guiding the child towards a system of socially established and intersubjectively agreed judgements.

Baldwin's genetic epistemology is *functionalist*. The notions of accommodation, assimilation and organic selection designate functional mechanisms for the elaboration and progressive adaption of behaviour to the environment. This environment *pre-exists* the organism (Baldwin, unlike Piaget, did not adopt a radical constructivist approach to knowledge and representation), but is also respondent to the actions of the organism. It is in the divergence between the intended or desired outcomes of behaviours—guided by what Baldwin calls 'interests'—and their actual outcomes, as either directly experienced through the 'control' exerted by reality, or as socially signified in the form of a 'mediated control' exerted by other subjects, that accommodatory representations, and ultimately the linguistic concepts of predication and implication, have their origin. Thus, it is the contradiction between need/desire and reality, and the gap between actual and imagined adequation, which stands at the heart of Baldwin's dialectic.

Baldwin's genetic epistemology is also as much social as it is psychobiological in its orientation. In this respect, and in the importance it accords to language and communication, it has evident affinities with Mead's and Vygotsky's theories, and is appreciably different in emphasis from Piaget's treatment of the same themes, to which I now turn.

Accommodation and assimilation (2): Piaget

As Russell (1978: 6) points out, although Baldwin did more than merely to prefigure Piaget's theory, Piaget equally did more than to simply extend Baldwin's theory. In fact, the two accounts differ more than is at first suggested by their use of the same terminology. In the first place, Piaget inverted the priority assigned by Baldwin to accommodation over assimilation. Second, he introduced an additional key term, *equilibration*. Third, in contrast to Baldwin's functionalism, Piaget's theory is essentially structuralist; his three functional concepts, of which the most important is equilibration, are best understood as constructs intended to account for the problem of structural stability and transformation. Fourth, Piaget emphasized the co-ordination of action *within the individual* as the basis of intelligence, and his main theoretical propositions concerned the structural elaboration of such co-ordinations through the stages of sensori-motor and operative intelligence (see Chapter 3).

As a consequence of these differences, Piaget's theory is a more radical epigeneticism than Baldwin's. Whereas Baldwin implicity accepted the *givenness* of the environment, both bio-physical and social, in relation to the subject, Piaget emphasized the literal *construction* of reality by the child. This does not mean that Piaget considered himself to be an 'idealist', for he steadfastly insisted that the constructions of intelligence were *necessary*, though not *a priori*, to use his own (rather questionable) terminology. The necessity of these constructions (or co-ordinations) resides, according to Piaget, in their double aspect as, on the one hand, biological adaptations, and on the other, as epistemological universals. As he put it, the co-ordinations are 'the necessary result of psychogenetic constructions yet conform to a timeless and general standard' (Piaget, 1977b: 25).

The co-ordinations constituting cognitive systems are viewed as literal extensions of the biological autoregulative mechanisms (Piaget frequently cites the concepts of homeostasis and homeorhesis) enabling the organism to maintain itself in dynamic equilibrium in relation to its environment. The co-ordinations thus function as assimilatory schemata, analogous to the behavioural and physiological systems which enable the organisms to assimilate food; Piaget frequently uses the metaphor of 'alimentation' when speaking of cognitive activity. All intelligence presupposes and depends upon assimilation, by which alone the structure of reality can be apprehended, and upon which perception itself depends.

Accommodation, however, is also implied by the very process of assimilation: for every instance of adaptive behaviour or cognition must be precisely fitted to the particular object, event or logico-mathematical structure to which it is directed. In order to grasp an object, the kinematic structure and dynamic configuration of the hand and arm musculature must anticipate the shape and position of the object: and such accommodations extend beyond the sensori-motor to the operations of logico-mathematical reasoning. Adaptation, for Piaget, is brought about by a process of equilibration between assimilation and accommodation. Accommodation, then, is for Piaget an anticipatory and corrective concept which is dialectically necessitated by every assimilative act. In this conception, he diverges from Baldwin, for whom accommodation is an essentially reactive process necessitated by the failure of existing accommodatory capacities to overcome object resistances.

Thus assimilation for Piaget plays the more basic role. Insofar as assimilation implies accommodation in every instance, it is both more active than it was for Baldwin, who saw it as an essentially passive registration; and more progressive: it is the 'disequilibrations' brought about by the inadequation of assimilation, and *not* the failure of accommodation, that necessitate and provoke re-equilibration at a higher level. Piaget, in his later writings, was very clear about this:

> We must distinguish two important categories of disturbances. The first includes those which are opposed to accommodations: resistances of objects, obstacles to reciprocal assimilations of schemes or subsystems, etc. In short, these are the reasons for failures or errors of which the subject becomes more or less aware; the corresponding regulations include negative feedback. The second category of disturbance, *the source of nonbalance*, consists of gaps which leave requirements unfulfilled and are expressed by the insufficiency of a scheme ... The gap, functioning as a disturbance, is therefore always defined by an already activated scheme of assimilation, and the corresponding regulation then includes a positive feedback which prolongs the activity of the scheme (pp. 1–19). The object not yet assimilated and not immediately capable of being assimilated constitutes an obstacle ... and a new accommodation is then necessary. But as assimilation and accommodation constitute two poles, and not two distinct behaviours, it is clear that the new assimilation plays the construction role, and the new accommodation that of compensation (Piaget, 1977b: 1–19, 39; emphasis added).

In summary, Piaget's constructivist epigeneticism is radically endogenous and subject-oriented, but its criteria of objectivity are timeless formal abstractions. In Baldwin's functionalist epigeneticism, on the other hand, overall systemic development is driven by endogenous processes awakened by exogenous changes (themselves brought about in part by the subject), and the criteria of objectivity are those of socially negotiated, intersubjective agreement.

Adaptation, affordance and representation

It is tempting to suggest that the concept of affordance, as proposed by Gibson, constitutes a reciprocal construct, in terms of the environmental niche, to that of assimilation, as proposed by Piaget, in terms of organismic adaptation. For it is in virtue of its affordances that an object offers itself to organismic assimilation; and to the extent that the object's affordances fail to correspond to the assimilatory schemes of the organism—that is, the object resists assimilation—then the organism must perforce accommodate its schemes and actions to the 'new' properties afforded—that is, extend its assimilatory capacities.

Such an interpretation receives support from the extension of Gibson's theory by Shaw, Turvey and Mace (1982), who propose an

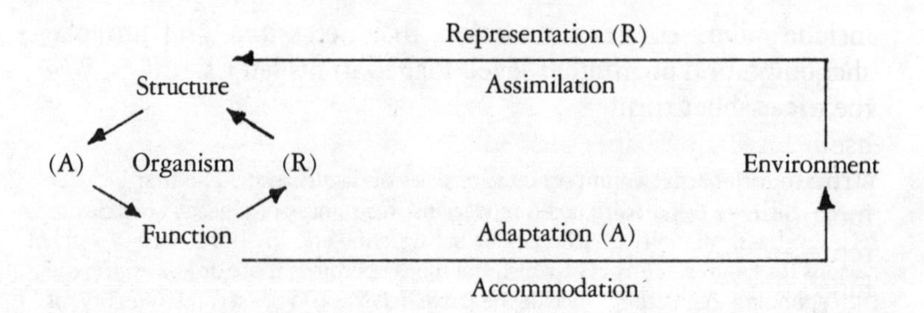

Figure 4.3: Representation and adaptation in an epigenetic naturalist framework

organismic counterpart of environmental affordance in the concept of 'effectivity'. The 'effectivities' of the organism, on this account, would correspond roughly to those assimilatory schemes adapted to the affordances of the niche.

The human environment, however, as I have repeatedly emphasized, has always-and-already been *intentionally shaped* by previous generations of human agents into a material and signifying culture, consisting of artefacts and semiotic systems embedded within social practices and institutions. Thus, the physical environment of the human infant is *meaningful in its material structure, representing* human consciousness and intentionality. Representation, in the epigenetic viewpoint, is no longer to be seen as 'merely' mental: consciousness and knowledge are 'inscribed', not just in brains and nervous systems, but also in artefacts, institutions, practices, symbols, utterances and languages. Representation, like behaviour, extends beyond the boundary of the individual organism. The human infant, in development, is engaged in an accommodatory effort after meaning, whereby culture and representation is assimilated and transformed by every generation.

Representation, from this viewpoint, is the structural realization of adaptive action, and adaptation (adequation) is the active engagement of the subject in the use and acquisition of representational systems in pragmatic contexts. Thus, the circuit of assimilation and accommodation is also the circuit of representation and adaptation. In Fig. 4.3, the directionality of the arrows designates the directionality of shaping processes: through adaptive functioning, the organism shapes the environment, including its representational structure; the environment also shapes the organism through being represented in the structure of the organism. Equally, the structure and function of the organism itself mutually shape each other. Just as the environment is structurally represented in the organism, so too does the adaptive functioning of the organism in the environment acquire a material representation in organismic structure; and just as organismic function is adapted to the environment, so too is organismic structure adapted to organismic function.

Representation and adaptation are regarded here as 'equilibrated' or more or less stable products of a *circuit* of accommodation and assimilation, a circuit which is the dynamic aspect of structural representation/functional adaptation. Figure 4.3 can therefore be seen as an attempt to reconcile Baldwin's functionalism with Piaget's structuralism. No distinction is drawn between 'cognitive' and 'morphological' structure: it is assumed that the organism progressively 're-represents', neurologically, systems of representation existing in the environment. Equally, behavioural and mental function are assumed to be aspects of a continuous, asymptotic adaptive engagement with a reality which is representational in its material structure. No assumptions are made about the relative priorities to accord to assimilation and accommodation. As a hypothesis, one might suppose that functional accommodations to augmented environmental (including social) complexity, especially when 'reflected back' by other subjects, resist structural assimilation, and thus provoke the kind of internal disequilibration that Piaget proposed.

Neural networks in epigenetic development

Neuroscientific research in the last few years has produced empirical findings and formal models regarding brain function and development which seem to offer a plausible biological basis for the socio-epigenetic theory which I have outlined. The models which I summarize in this section derive from two sources. The first is work in 'core' neuroscience, involving general issues in the development and functioning of the central nervous system, and some specific issues in the neuropsychology of memory (Changeux, 1985; Edelman, 1981; see also Rosenfield, 1986, for a detailed introduction to Edelman's theory). The second is work in cognitive science, involving the computational modelling of 'neural-like' networks in terms of Parallel Distributed Processing (PDP) (Rumelhart *et al.*, 1986).

Epigenesis through selective stablization

A central problem for theories of the development of the human brain and nervous system—in particular the development of the neo-cortex—is how the 10^{10} to 10^{12} neurons each come to have the very large number (upwards of 50,000 in some cases) of synaptic connections with other neurons (totalling 10^{15} to 10^{16}), which together form the basis for the neuropsychological activity of the brain; and, ultimately, for cognition, affect, perception and behaviour. Widely accepted metaphors such as 'imprinting' and 'etching' seem to suggest that the *environment* is somehow directly responsible for the growth patterns of synaptic connections, through a process of 'instruction' which leaves behind fixed 'traces' or copies. This assumption underlies the traditional empiricist views of learning, and seems (as many authors, including Chomsky, 1980, have noted) to stand in contradiction to the general Darwinian principle that development and evolution occur, not through Lamarckian 'direct action' or instruction, but through a process of *selection*. Chomsky (1980: 14) indeed cites the immunologist Jerne (1967) to the effect that 'Looking back into the history of biology, it appears that wherever a phenomenon resembles learning, an instructive theory was first proposed to account for the underlying mechanisms. In every case this was later replaced by a selective theory.' In previous chapters, I have emphasised the philosophical inadequacy of the 'copy' theory of perception and cognition; to these criticisms we may now add, then, the further criticism that the copy theory (and, by implication, empiricist and associationist psychology) is out of tune with the main thrust of evolutionary biological thinking during this century.

This argument is, at first sight, grist to the mill of the neo-rationalist, who will argue that what appears to be 'learning' and 'development' is *actually selection* from among genetically pre-programmed (hardwired) connections. Such a theory is advanced by, for example, Mehler (1972), who suggests that ontogenetic cognitive development may be a process of impoverishment of rich innate dispositions, whereby initial global processes are replaced by later more specific processes. This approach also characterizes neo-rationalist accounts of language acquisition (Atkinson, 1982; Pinker, 1979; Wexler, 1982), which postulate *constraints on learnability*, reflecting properties of UG, coupled with a restriction of the role of input (at least where syntax is concerned) to the *setting of parameters*.

The hypothesis that learning proceeds through selection rather than through instruction is supported by current epigenetic models, in which 'pre-existing connections are selected by activity or "experience" without inducing any synthesis of new molecular species or structures' (Changeux, 1985: 229). Such models depart from neo-rationalist, nativist conceptions, however, inasmuch as neurological structure is not strictly genetically pre-programmed. The neurobiological analogue to the neo-rationalist thesis would consist in a model of neurological development in which specific genetic markers determine, for each neuron, the synaptic connections which it will make.[34] In such an account, neuro-development would be genetically determined to the extent that, for example, identical twins might be assumed to have structurally identical brains, in terms of patterns of connections.

This does not appear, however, to be so: the process of neural development is not so strictly deterministic as such a view would suggest, at least at the level of the construction of individual synapses. Rather, the development of brain connectivity is best seen in terms of the *selective stabilization* of a set of initially labile states, in the course of epigenetic (non-deterministic) growth. This process is under genetic control at a higher level, but such control allows for (indeed inevitably produces) *variability* in the developmental processes of even genetically identical individuals. This is how Changeux (1985: 227–228) expresses 'The theory of epigenesis by selective stabilization of synapses':

> [It has been established that]
>
> 1. The principle features of anatomical and functional organization of the nervous system are preserved from one generation to another and are subject to the determinism of a set of genes that make up what I have called the *genetic envelope*. This envelope controls the division, migration, and differentiation of nerve cells; the behaviour of the growth cone; the formation of widespread connections; and the onset of spontaneous activity. It also determines rules governing the assembly of molecules in the synapse and the evolution of this connecting link.

2. A phenotypic variability is found in the adult organization of isogenic individuals, and its degree increases from invertebrates to vertebrates, including humans, parallel to the increase in brain complexity.

3. During development, once the last division of neurons has taken place, axonal and dendritic trees branch and spread exuberantly. At this critical stage, there is redundancy, but also maximal diversity in the connections of the network. This redundancy is temporary. Regressive phenomena rapidly intervene. Neurons die, and a considerable portion of the dendritic and axonal branches are "pruned". Many active synapses disappear.

4. Impulses travel through the neuronal network even at a very early stage of its formation. They begin spontaneously, but are later evoked by the interaction of the newborn with its environment...

[It may further be hypothesized that]

The evolution of the connective state of each synaptic contact is governed by the overall message of signals received by the cell on which it terminates. In other words, the activity of the *post*synaptic cell regulates the stability of the synapse in a *retrograde* manner ... Epigenetic development of neuronal singularities is controlled by the activity of the developing network. It commands the selective stabilization of a particular set of synaptic contacts from within the total set present during the stage of maximum diversity.

The selection process, in both Changeux's and Edelman's models, operates on populations or groups of neurons, which are subject to control by a common 'genetic envelope' which specifies the initial group (or groups) within which connections are to be established, without specifying the eventual 'singularity'—the specific constellation, for each neuron, of synaptic connections which results from the process of selective stabilization. As Changeux (1985: 248) points out, 'The laying down of "redundant" and "variable" neuronal or synaptic topologies—the substrate of epigenesis—costs much less in genetic information than would a point-by-point coding of the diverse neuronal singularities found in the adult. The genes that make up the genetic envelope, in particular those that determine the rules of growth and stabilization of synapses, can be shared by all neurons in the same category, perhaps by several categories of neuron.' A consequence of this process is that the set of singularities constituting a given mature organic individual is *unique*, although the neuro-psychological activities that this set of singularities supports may be assumed to be shared with other individuals. This leads to a model in which 'the same afferent message may stabilize different arrangements of connections, which nevertheless result in the same input–ouput relationships ... different learning inputs may produce *different* connective organizations and neuronal functioning abilities but the *same* behavioural capacity (Changeux, 1985: 228, 247).[35]

The process of selective stabilization, according to Changeux, is neither 'once-off' nor even: '[Synaptic connections] proliferate in successive waves from birth to puberty in man. With each wave, there is transient redundancy and selective stabilization. This causes a series of critical periods when activity exercises a regulatory effect ... to learn is to stabilize pre-established synaptic connections, and to *eliminate* the surplus' (1985: 249). At any given stage during this epigenetic process, the current state of a group of neurons will be determined partly by the spontaneous growth of new connections; partly by the past history of selective stabilization; partly by new activity states arising from experience; and partly by concurrent and previous developmental processes in other groups of neurons. It is these characteristics of the process of selective stabilization which together make up a model in which 'brain function, like structure, also depends on context and history and not on localized functions and fixed memories' (Rosenfield, 1986: 24).

These features of the theory of selective stabilization recall Luria's discussion of the nature of 'functional systems': 'that every movement has the character of a complex functional system and that the elements performing it may be interchangeable in character is clear because the same result can be achieved by totally different methods ... Although this 'systemic' structure is characteristic of relatively simple behavioural acts, it is immeasurably more characteristic of more complex forms of mental activity ... mental functions, as complex functional systems, cannot be localized in narrow zones of the cortex or in isolated cell groups, but must be *organized in systems of concertedly working zones, each of which performs its role in the complex functional system*, and which may be located in completely different and often far distant areas of the brain' (Luria, 1973: 29, 31).

The epigenetic models proposed by Changeux and Edelman go considerably further than Luria's concept of 'functional system', however, both in the degree to which the developmental mechanism is specified[36] and in the particular concepts of neurocognitive representation which they propose. In Changeux's model, an initially labile system is selectively stabilized to form a durable pattern of connections. The augmentation in development of the durability of the set of connections constituting a singularity is realized also at the level of the individual synapse: 'Whereas the neurons of a human brain can survive for more than a hundred years, the life of synaptic molecules is much shorter. In the adult neuromuscular junction, the half-life ... is about eleven days. But those molecules that disappear are immediately replaced by other, newly synthesized ones. The molecular architecture of the adult synapse is constantly renewed so that its organization remains stable. In the embryonic muscle fiber, the situation is ... more fluid. The half-life of the receptor is very short, only eighteen to twenty hours. The embryonic receptor is very labile' (p. 224).

In this model, the epigenetic developmental history of the organism is not only formative in the construction of the singularity, but this history (including the history of organism–environment transactions) achieves a *representation* in the structure of neural connections: '[neuro-cognitive] representations are built up by the activation of neurons, whose dispersion through multiple cortical areas determines the figurative or abstract character of the representation. A mental object is by definition a transient event. It is dynamic and fleeting, lasting only fractions of a second. The singularities of the neurons that form it, however, are much more stable; they are built up during development by mechanisms involving internal genetic expressions and regulations stemming from a chain of reciprocal interactions with the environment. Thus the epigenetic component of neuronal singularities itself constitutes a "representation", written in the "wiring" between the nerve cells' (Changeux, 1985: 227).

It is precisely this conception of representational development, in which assimilation is viewed as the structural realization of accommodatory adaptations, which was proposed in the previous section, and which now appears to gain support from contemporary neuroscience.[37] To continue, however; the model proposed by Edelman and his colleagues proposes a specific theory of memory and perception in which neuronal groups (interconnected singularities, in Changeux's terms) are structured at a higher level in the form of 'maps'; these maps are in turn interconnected in such a way that the products of inter-map communications (in the form of 're-entrant connections') may be seen as flexible and context-sensitive mechanisms for categorization and recategorization. In Edelman's model, updating and perceptual re-categorization proceed automatically on the basis of the changing structure of incoming information. The theory therefore provides a model for a process of recognitory assimilation which is sensitive to a dynamically structured context of *background*.

A further implication of Edelman's model is that the memory 'trace' underlying a given category (or rather, a given instance of the process of categorization) is not a fixed template (copy), but a configurational state which is 'read' in relation to co-ordinate states. Thus, memory is constructive (or reconstructive), rather than passive. Not only is this view compatible with the overall epigenetic approach advocated here, but it is generally in line with the theories of memory proposed by both Bartlett (1932) and Piaget and Inhelder (1973), as well as more recently by Schank (1982). Most interestingly, however, the mechanisms proposed by both Changeux and Edelman appear to share important properties with the 'neo-connectionist' models of representation developed by Rumelhart and the PDP research group, to which we now turn.

Neo-connectionist computational modelling

As we have seen, the meta-theoretical stance underlying the kinds of computational models advocated by neo-rationalist theories—and indeed by almost all cognitivist accounts—is based upon the assumption that the specification of Turing-modellable 'effective procedures', of a serial-symbolic (von Neumann) nature, constitutes the strategic and heuristic goal of cognitive science. Rumelhart *et al.* (1986) advocate, in contrast, a theory of representation and cognitive processes based upon Parallel Distributed Processing. This neo-connectionist approach is based in part upon ideas in computational design for parallel processing architectures—which, it is widely believed, will form the basis of a new generation of computer hardware—and in part upon a research strategy in cognitive science which takes its inspiration from the modelling of neural networks. As various authors have noted (e.g. Gardner, 1987), this work may be seen as in some respects a return to the problematic of an earlier period in cognitive science, exemplified by the work of McCulloch and Pitts (1943), in which the modelling of the known structure of the brain and nervous system, rather than the axiomatization of formal domains, is seen as the fundamental goal.

Rumelhart *et al.* insist that their work is better seen as 'neurally inspired' than as literal modelling. In this respect, PDP models fall somewhere in between Modelling_1 and Modelling_2, as defined above (p. 116). Nevertheless, given that PDP models begin from the assumption of constraints on the realization of cognitive processes in terms of human brains, they represent a decisive break from the free-floating formalisms of the contingency assumption.[38] Interestingly, however, the high degree of constraint imposed upon PDP models in the interests of neural veridicality eventuates, in many of the implementations so far studied, in a type of computational 'power' which diverges widely from the conventional understanding of the term; and which Rumelhart *et al.* discuss in terms of 'emergent properties'.

The type of 'power' characterizing 'emergentist' machines may informally be compared with the conventional notion of computational power in relation to an issue addressed earlier, in the discussion of 'input analyser modularity' (p. 120): parsing. Rumelhart and McClelland (1986: 119) point out that, while it is possible to implement recursive engines by means of PDP representations, 'We have not dwelt on PDP implementations of [such] engines because we do not agree with those who would argue that such capabilities are of the essence of human computation ... the human ability to use semantic and pragmatic contextual information to facilitate comprehension far exceeds that of any existing sentence processing machine we know of. What is needed, then, is not a mechanism for flawless and effortless processing of center-embedded constructions. Compilers of computer languages generally provide such facilities, and they are powerful tools, but they have not demonstrated themselves sufficient for processing natural language. What is needed instead is a parser built from the kind of mechanism which facilitates the simultaneous consideration of large numbers of mutual and interdependent constraints.' They suggest that it is this kind of computational task for which PDP models are suitable.

Hinton, McClelland and Rumelhart demonstrate, further, that distributed knowledge representations, in which 'Each entity is represented by a pattern of activity distributed over many computing elements, and each computing element is involved in representing many different entities' (p. 77) not only possess significant advantages over local (point to point) representations in terms of modelling content-addressable memory[39] and 'best fit' rule selection, but also exhibit the same kind of 'constructive' or 'reconstructive' processing properties as are implied by Edelman's epigenetic model of memory. Context, history and co-activity influence recognition processes based upon stabilized activity patterns which, rather than being 'copies' filed at specific locations, are 'virtual' or 'latent' sets of relationships until activated and stabilized by input: 'patterns which are not active do not exist anywhere. They can be re-created because the connection strengths between units have been changed appropriately, but each connection strength is involved in storing many patterns, so it is impossible to point to a particular place where the memory for a particular item is stored' (p. 80).

Rumelhart, Smolensky, McClelland and Hinton develop these notions into a general theory of recognition, based on constraint satisfaction networks. The satisfaction of constraints proceeds in terms of a process of optimization of the system to the input, which Rumelhart *et al.* call 'relaxation'. These network properties may be represented in terms of a 'goodness-of-fit landscape' which bears striking similarities to the concepts of 'epigenetic landscape', 'chreods' and 'canalizations' which Waddington (1975, 1977) employs to represent developmental and evolutionary processes. The 'constraint satisfaction network' theory of recognition also corresponds in many ways to the concept of 'recognitory assimilation' proposed by Piaget, as is clear from Rumelhart *et al.*'s discussion of the concept of 'schema'.

As they point out, a major problem for most (local) notations for representing schemas, scripts or frames (Minsky, 1975; Bobrow and Norman, 1975; Schank and Abelson, 1977) consists in the appropriate assignment of 'default values'. In most implementations, such assignments are context-independent, although, ideally, constraints upon possible values should be considered as multivariate distributions capturing inter-dependencies between 'slot-filling' values. Less formally, Rumelhart *et al.* express the problem as follows: 'On the one hand, schemata are the structure of the mind. On the other hand, schemata must be sufficiently malleable to fit around most everything' (vol. 2, p. 20). The most important aspect, for our purposes, of the solution proposed by the PDP model is that 'schemata' are no longer conceived in terms of fixed categories, but are the dynamic products of organism–environment transactions, which deliver recognitory assignments from out of the 'foregrounding' of an input in relation to the current state of the network as a whole. As Rumelhart *et al.* put it, 'There is no representational object which is a schema. Rather, schemata emerge at the moment they are needed from the interaction of large numbers of much simpler elements working in concert with one another. Schemata are not explicit entities, but rather are implicit in our knowledge and are created by the very environment that they are trying to interpret—as it is interpreting them' (vol. 2, p. 20).

In place of the neo-rationalist mechanism of *selection of the environment by the system*, the PDP variant of the schema offers a mechanism of *mutual selection*, as part of a construction process in which the 'givens' of both system and environment achieve representation *only through the interactions between them.* Within an epigenetic perspective, such transitory, 'on-line' representations acquire a more permanent representational status in development; not as fixed 'elements', but as structural selections from among an array of potential configurations. Each such selection may be seen as both potentiating further processes of assimilatory growth, and constraining the array of possible directions of growth.

Multiple representations encoding related information (such as the semantic value of a discursive concept, and information about its denotation) may serve, in such a model, to modulate the recognitory processes of each other, without being reduplicated as fixed information. This would constitute a computational equivalent to the relationship between 'real world knowledge' and 'linguistic knowledge', which I described in Chapter 2 as being a 'partial and permeable' distinction, and which I further investigate in terms of context, background, presupposition and canonicality in Chapter 5.

Problems and perspectives

Of the many issues arising from the previous discussion, perhaps the most important is that of *reductionism*, and in particular whether neural-network inspired theories necessarily imply (a) a return to crudely associationist theories of learning; and (b) a physicalist ontology in which, to put it simply, there are 'only brains', and mind is a mere epiphenomenon. Both these views have been the target of repeated and cogent critiques by neo-rationalists, the force of which I do not wish either to deny or to dispute. My intention is rather to suggest that neural network theories do *not* imply either associationism, or reductionist physicalism, and to argue for a non-reductionist interpretation of neural network theories from within a socio-naturalistic, epigenetic perspective.

I want to start by suggesting that the problem of reductionism can be approached in two different ways, namely in terms of *mechanism* and in terms of *epistemology*. I do not mean to suggest that there is a strict division between the types of questions that occur in relation to mechanistic and in relation to epistemological concerns; obviously the two are closely related. However, the distinction may help to bring some conceptual order to a complex field of problems. The principal difference between 'mechanistic' and 'epistemological' problems related to neural network models is that the former are susceptible (in principle) to solution on empirical grounds, whereas the latter, in all likelihood, are not.[40]

To begin with the issues of mechanism: the central questions here concern the extent to which epigenetic, neural network models are compatible with empirical evidence and current hypotheses regarding human cognitive and communicative development.[41] The two particular questions that I wish to address here are (a) modularity and re-organization; (b) grammar and psychological reality. The latter question will serve as a bridge to the epistemological issues.

Modularity and re-organizational processes

I argued earlier in this chapter that the 'input analyzer' version of the modularity thesis advanced by Fodor (1983) is theoretically neutral with respect to the neo-rationalist paradigm. The question here is whether or not an epigenetic, neural-network theory can accommodate modularity. The first thing to be said is that Fodor's arguments for the modularity and 'encapsulation' of diverse cognitive processes are extremely well founded; much of the resonance with which his monograph has met derives from the way in which it meshes with what we know of developmental phenomena in various psychological domains. The most exemplary such domain is *language*, which is widely accepted to constitute (in the well-known expression of Karmiloff-Smith, 1979) a *problem space in its own right*, whose acquisition is difficult to explain by recourse only to general and content-independent principles of learning. This is not to say, of course, that language acquisition is *independent* of other aspects of cognitive development. Although the case for the absolute autonomy of language acquisition has frequently been argued or assumed by neo-rationalist theorists, there is ample evidence against it (see for example Bloom, Lifter and Broughton, 1985). The hypothesis that is under discussion here is a 'weak' (but substantive) modularity hypothesis, and not a 'strong' (absolute) one.

Since this is not an empirical review, I shall simply take as given the wealth of evidence supporting the view that the acquisition of language exhibits certain processes—notably processes of rule construction and reconstruction and systemic re-organization—which proceed along paths which are *internally* determined by the properties of the developing language (sub-)system(s), and not *externally* determined, either by other cognitive systems and structures, or by general functional considerations (Karmiloff-Smith, 1987). Again, this does not mean that other cognitive systems, or functional determinations, are *irrelevant* to or *uninfluential upon* language acquisition (Paprotté and Sinha, 1987); rather, such external determinations are not exhaustively explanatory of language acquisition and development, which must therefore be considered to be a relatively autonomous, or weakly modularized, process: or, rather an interacting complex of weakly modularized processes.

Now, the question is *not* whether, given sufficient initial information, a network system can model such processes, since the work of Rumelhart and the PDP group seems convincingly to suggest (in relation, for example, to the acquisition of the English tense system) that it can. Rather, the question is whether the necessary initial information is *specific to language*, in the particular sense implied by UG. This question is currently unanswerable on empirical grounds, and in all likelihood will remain so for some considerable time. What I wish to suggest, though, is that it is at least logically possible to suppose that the epigenetic development of particular cognitive subsystems is *specialized* to particular informational natural kinds (including natural language and its sub-components), while not being endowed with initial information specifying the *structure* of the particular informational natural kind, in the generative sense. What would be required, on this view, is the genetic specification of the variables constituting a domain of *recognitory assimilation*, rather than the genetic specification of the variables 'parameterized' by a construction process. On this account, within-domain processes of reciprocal accommodations, assimilatory restructurings and equilibrations would eventuate in emergently 'self-organizing' systems.

Such a process might appear to an observer like the parameterizing of values, but the explanation for it would differ from the neo-rationalist account in several important respects. First, the data for the construction process—in the case of language acquisition—would be *the inputs of actual discourse*—not such inputs 'filtered' through UG. Second, the genetic specification of a domain would be in terms, not of the computational system constituting the formal description of the domain, but of the computational system necessary to recognize foregrounded *instances* of the domain in question, and to process them to the extent necessary to 'assimilate' or evaluate them against the background state of the total system (including other 'modules').

At a neurobiological level, this would imply that the genetic 'envelope' governing the growth of language and speech centres would specify both the topological regions within which connections are selectively stabilized, and the global structure of connectedness to other regions, which would also be subject to selective stabilization.[42] It might be hypothesized: (1) that the genetically specified informational specialization of language regions consists in attunement to modality-specific serial segmentation of stimuli, and to the rapid selective stabilization of combinatorial rules operating on the planes of both succession and simultaneity; and (2) that the privileging of connections to efferent motor systems governing vocalization and gesture would also be specified by the genetic envelope. Epigenetic processes, open to the informational environment and subject both to spontaneous waves of connective multiplication, and to selective stabilization of connections and connection strengths, would account for the further development of the system.

It is at least conceivable that such a model could offer sufficient structural complexity and developmental capacity to explain certain well-attested phenomena in language and cognitive development, to which traditional, incremental learning theories are ill suited. Such phenomena have been described in terms of 'U-shaped curves', 'apparent regressions' and 'cognitive pendulum swings' (Nelson and Nelson, 1978), and are observable in a variety of different cognitive domains in the form of developmental shifts in *strategies* of use and acquisition. These strategic shifts may involve, both within and between cognitive and linguistic domains, shifts in the *level of determination* (semantic, syntactic, morphological) of processes of re-organization and of integration (Bowerman, 1982; Sinha and Carabine, 1981).

Not only language, but other developmental domains frequently display complex and apparently inconsistent inter-relationships between continuity and discontinuity; patterns of global irreversibility and local reversibility; disintegration—differentiation—reintegration cycles; and alternations between 'penetrability' and 'impenetrability' to other concurrent processes. The explanation for these phenomena, in neural network terms, would be sought in a combination of spontaneous augmentations in the number of connections (with supervening selective stabilization processes); and changes in weightings associated with specific connective pathways. At a more general level, such processes may be defined in terms of the theory of self-organizing systems (Prigogine and Stengers, 1984). Thelen, Kelso and Fogel (1987) propose, for example, an emergentist, self-organizing system-theoretic account of motor development, in which 'new forms ... may arise as discontinuous phase shifts, although the underlying metric may be continuously scaled. Thus, a small change in a crucial parameter may be amplified and shift the entire system into a new, and qualitatively different, mode. In development, these may be recognized as stages' (p. 59).

It cannot yet be claimed that fully specified epigenetic network theories of development (either in language or in other domains) are currently available.[43] It seems reasonable, however, to suggest that, at the mechanistic level, such theories, far from being reductionist, can offer powerful explanations for phenomena which are simply inexplicable in traditional, associationist terms. An epigenetic approach is also able to preserve the insights of the modularity thesis, while recasting it in developmental terms: modularity is seen as a largely self-specifying process, which is initially only weakly determined, but which progressively reconstitutes its informational environment (domain) in the direction of increasing autonomy and specialization. This is essentially the notion that Luria (1973) denotes as 'automatization', and which has 'encapsulation' as its end state.

Encapsulation is not, however, directly genetically specified, but is the result of epigenetic interactions in which impulses for reorganization derive from both internal and external sources. Modularization, on this account, is a psychobiological consequence of the epigenetic progession from plasticity and lability to specialization and stability; and from (selective) penetrability to (relative) informational encapsulation. The psychobiological process of modularization, however, as I shall argue, should not be conceived in terms of the acquisition or construction of *internal representations* corresponding to objects such as grammars.

Psychological reality revisited

The view which has dominated psycholinguistic theory and research for the past three decades is that the 'object' of the language acquisition process is identical with the 'object' of linguistic theory: that is, *competence* in one's native language. The view which I propose, on the contrary, rejects the notion that the mental representations constituting the psychological capacities of a speaking and comprehending subject are to be identified with a formal object such as a grammar. In traditional terms, epigenetic, socio-naturalistic theories would thus be considered as 'performance models'. In many ways, however, this usage is misleading, since the entire thrust of the approach is to *dissolve* the competence–performance distinction, at least as currently understood. This should not, it should be noted, be taken to imply that Generative Grammars (for example) are incorrect *qua* linguistic theories. Rather, it implies that such grammars are not, and cannot in principle be, theories of the psychological representational systems underlying the acquisition and use of language.[44]

Language, in the sense in which it is represented by linguistic theory, should rather be seen—as it was by Saussure—as an objective social institution, which is exterior to, and developmentally prior to, the psychological subject. As McNeill (1979: 293) puts it, 'Grammars ... refer to real structures, though not to psychologically real structures in the processing sense ... a grammar is a description of our *knowledge of a social institution*—the language—and because of this basis in social or institutional reality, rather than in cognitive functioning, grammars and psychological processes have no more than the loose relationships they appear, in fact, to have. The role of grammar during speech programming is analogous to the role of other social institutions during individual behaviour. This role is to define and evaluate the behaviour of individuals. It is not to cause the behaviour.'

The representations which are rendered *explicit* in formal descriptions or devices such as grammars play an *implicit* role in psychological mechanisms of discourse production, comprehension and acquisition, but not a role which can be defined in terms of generation and interpretation. Rather, grammars provide criteria to which speakers and hearers implicitly refer in evaluating utterances in discourse. These criteria are available to speakers and hearers only in the form of the representations which actuate them—in Chomskian terms, only as 'surface structures'. In place, however, of the neo-rationalist appeal to formal *knowledge* of deep structure (or UG), a socio-naturalistic, epigenetic approach proposes a process of *adaptive adequation* to the conditions on intelligible representation (Chapter 2). The completion of these conditions can occur only on an intersubjective plane. Thus, the appropriation of evaluative criteria ('grammars') by individual subjects consists not in the internalization of an abstract object (the grammar), but in the construction and mastery of procedures for *acting intersubjectively* within the system of supports and constraints offered by the institution of language.

Children, in learning language, are engaged in negotiating their place—or places—in a complex web of social and communicative practices, some of which are linguistic practices. Linguistic practices are supported and constrained by the institution we call 'language', and this institution is actuated in utterances and texts. Utterances and texts, however, actuate a variety of other institutions too—in effect, all those representational and signifying practices which together constitute culture. Conversely, human practices (discursive and non-discursive) achieve representation in a variety of material forms—not only in signifying systems, but also, for example, in the artefacts which are the products of human labour. What the child learns, and what all subjects employ, is a system (or set of systems) of representation, in which many aspects of context are co-present, and integrated at multiple levels as social, psychological and neurological processes. The nature of such representations, whether language as 'representing' cognitive processes, or the brain as 'representing' language, are such as to imply, not 'transcription', but *mutual functional adaptation*. The interdisciplinary relation between, say, neurology and linguistics, is then underpinned not by the identity of their theoretical objects, but by the *real functional interconnectedness* of the systemic articulations of the brain, as a representational system, and language, as a representational system.

The theoretical position which I have advocated is *materialist*; but it cannot be identified with any form of neurological or physicalist *reductionism*, since the materiality of mind, on this account, extends beyond neurological structures to encompass the totality of representational and signifying objects and systems which sustain social and communicative practices. This perspective does not at all deny that language (both as an institution and as a particular psychological domain) possesses unique characteristics as a 'problem space'. However, language-specific domains such as syntax are articulated in, and acquired in, a *context*, and are not (institutionally or psychologically) independent of, or closed to, that context (see, for example, de Lemos, 1981). The interplay between 'context' and 'representation', 'implicit' background and 'explicit' foreground, is characteristic of processes at every functional level from the neurological to the social-interactional. This interplay is examined in the next chapter, with reference to object representation and word meaning.

Notes

1. *The Science of Logic*. Cited in Lenin (1961).
2. *Love's Labour's Lost*, IV. ii.
3. See Langacker (1987: 28ff.) for a discussion of the 'exclusionary fallacy' in relation to linguistic theory.
4. Schilcher and Tennant (1984) offer an instructive, undogmatically interpreted account of contemporary neo-Darwinism which is quite consistent with the view that epigenetic theories are a part of, rather than in

opposition to, neo-Darwinian theory in the wider sense. It may be conceded then (as Schilcher and Tennant suggest with regard to a number of controversial issues in evolutionary theory) that epigenetic 'modifications' to neo-Darwinism do not necessarily affect its basic structure. In that sense, there is no question of a 'deep' paradigm difference. I should add, however, that Schilcher and Tennant's own interpretation of evolutionary epistemology differs in several important respects from the theory presented here.

5. See, for example, the debates in Piatelli-Palmarini (1980), which display every evidence of the dialogue of the deaf so characteristic of encounters between supporters of opposing paradigms.
6. In this chapter, I also use the terms 'cognitivist' and 'computational' to denote, respectively, the theoretical and methodological propositions which are programmatically fused in the modern neo-rationalist synthesis. 'Cognitivism', interpreted widely, consists in the proposition that intelligent behaviour can be explained only 'by appeal to internal cognitive processes, that is rational thought in a very broad sense' (Haugeland, 1978: 215). 'Computational' theories may be defined as those which view cognitive processes in terms of formal models of domains, which may only loosely be tied to particular implementations (or algorithmic representations, in the terms of Marr, 1982) of the computational theory. Marr's distinction between computational and algorithmic levels of theorizing maps onto the Chomskian competence-performance distinction which I express below in terms of the 'contingency assumption'.
7. It might be objected that the hegemony which I attribute to neo-rationalism within cognitive science is belied by the variety of work and theoretical positions encompassed by that interdiscipline (see, for example, the wide-ranging survey by Gardner, 1987). However, this is certainly not the way the neo-rationalists themselves see it—Fodor has famously and repeatedly asserted that the 'Establishment view' (cognitivism plus radical nativism) is 'the only ball game in town'.
8. The contingency assumption is absolutely basic to the modern neo-rationalist synthesis, and profoundly differentiates it from epigenetic and other naturalisms. It is expressed (as a 'type physicalist' variant of functionalism) in the following terms by Fodor (1981: 8–9): 'Given the sorts of things we want to say about having pains and believing P's, it seems to be at best just accidental, and at worst just false, that pains and beliefs are proprietary to creatures like us ... Whereas, what does seem to provide a natural domain for [cognitive] psychological theorizing is something like the set of (real and possible) information processing systems.'
9. I shall not try to place the term 'Artificial Intelligence' here, since it may be regarded more as denoting disciplinary affiliation than as specifying theoretical orientation.
10. Thus, Modelling_1 corresponds to the 'algorithmic' level, and Modelling_2 to the 'computational' level, according to Marr (1982).
11. In this respect the position of Johnson-Laird (1983), who stipulates that psychological theories should be couched in terms of 'effective procedures', is

no different, despite its terminological variations, from other computational approaches.

12. Given that the widely accepted account of Marr, 1982, further distinguishes 'algorithm' from 'implementation', the level which is addressed by 'computation' is in neo-rationalist theory actually at *two* levels of contingent remove from behaviour.
13. According to Gardner (1987), 1956 is in fact the most widely accepted date.
14. Strictly speaking, input analyzers are 'pre-perceptual', if perception is taken, with Fodor, to imply belief-fixation.
15. Fodor is careful to point out that Chomsky's own Cartesianism-plus-mental organs formulation of the modularity program may obscure the opposition which existed, historically, between Cartesian nativism and faculty theories (see ch. 1).
16. Bierwisch, 1987, neatly summarizes the current Chomskian position as follows: 'Recent developments in Generative Grammar have largely dispensed with grammatical rules in favour of general principles, varying with respect to a finite number of parameters. Under this assumption, UG is a modular system of principles with parameters to be fixed during language acquisition resulting in a system G' [i.e. grammar minus lexicon].'
17. I am aware, among other problematic issues raised here, of the difficulty involved in identifying a certain type of 'formal constraint' with (propositional) 'knowledge'. However, in defence of the argument that I have appropriated from Schilcher and Tennant, it should be said that it is exactly this identification which is proposed by the UG/LAD thesis.
18. In other words, it is not only not necessary to know what, or that, a stimulus *represents* to recognize it, it is also not necessary to assume the reduplication of limitlessly powerful formal descriptions of *classes* (formal domains) of stimuli in the recognitory devices which process them.
19. Langacker, in common with functional grammarians, therefore rejects the notion of the autonomy of syntax which motivates the necessity for the particular type of nativism espoused by Chomsky.
20. Fodor characterizes central processes as 'Quinean and isotropic', which properties I do not intend to define here; suffice it to say that, as I read it, Fodor's claim that central processing exhibits 'global' as opposed to 'local' computational properties *almost* equates to the claim that such central processing is not computational at all. Presumably, Fodor would maintain that global computations may be modelled by meaning postulates; but this still does not meet the kind of criticism expressed in the quotation from Searle in ch. 2.
21. In fact, there is nothing in the theoretical position which I am advancing which would preclude input analysers being computationally conceived in a general sense. If they were so conceived, they would be (in principle) instances of Modelling_1 above, not Modelling_2.
22. At the very least 'sub-doxastic states' (Stich, 1978) are implicated, which take as their objects non-empirical entities constituting abstract formal descriptions of domains such as 'the set of possible natural languages'.
23. I am employing the term 'mental model' here in a sense which is wider than

that of Johnson-Laird (1983).

24. The Chomskian account of language acquisition will be readily recognizable here.
25. In *The Modularity of Mind*, Fodor suggests an alternative solution to this problem; that is, that the outputs of input analysers are basic-level categorizations, in the sense of Rosch (1977). This proposal is certainly ontogenetically and phylogenetically well motivated; but it is not clear what, if any, relationship such outputs might have to the 'lexemes' of LOT. Of course, if LOT does not exist, but natural languages do, then one might appeal to experience of discourse and discourse contexts—but then that would be another kind of theory.
26. Fodor (1983: 9) revealingly remarks that 'Performance mechanisms do for Chomsky some of what the pineal gland was supposed to do for Descartes'—i.e. to provide a junction between 'mental' and 'material'.
27. See for different views Beaugrande, 1983; Boden, 1977, 1981; Dreyfus, 1972; Dennett, 1979; Karpatschof, 1982; Hofstadter, 1979; Plunkett and Larsen, 1988; Searle, 1980; Weizenbaum, 1976.
28. This is perhaps the right place to refer to—without attempting to summarize or to engage in any serious way with—Wilden's (1972) foundational critique of the notions of communication, structure and exchange as employed in the human sciences. To address this work adequately would probably require a book in itself.
29. For a historical account of Gibson's developing thought, including his gradual break with behaviourism, see Costall, 1981; for a discussion of varieties of 'ecological psychology', see Cutting, 1982.
30. It is interesting to note the similarities between Gibson's proposals and those of Goethe (1970) [1840] regarding the perception of colour. Goethe, too, opposed Newtonian optics with 'naturalistic' optics: 'The eye may be said to owe its existence to light, which calls forth, as it were, a sense that is akin to itself; the eye in short, is formed with reference to light, to be fit for the action of light ... [the aim of Goethe's theory being] to rescue the attractive subject of the doctrine of colours from the atomistic restriction and isolation in which it has been banished in order to restore it to the general dynamic flow of life and action which the present age loves to recognise in nature' (Goethe, 1970: lii–liv).
31. There is perhaps no intrinsic reason why formal or computational models should not be developed for direct perception processes; but if cognitive science were to proceed in such a direction, it would necessarily have to leave behind some of its long-cherished assumptions: not least those underlying the neo-rationalist synthesis.
32. See, however, de Gelder, 1987, for a different view regarding the compatibility of direct perception and representational theories.
33. An extensive discussion of Baldwin's theory and its relation to that of Piaget is to be found in Russell, 1978, upon which this account is partly based.
34. According to Rosenfield, 1986, such an account has been proposed by Sperry.
35. Changeux adds that this is despite the 'deterministic' nature of the model; by which he means not that the end-point of development is predetermined,

but that mechanistic-deterministic processes stochastically interact in development.

36. In the case of Edelman's work, an important role is accorded to the biochemical structures referred to as 'cell adhesion molecules'; this aspect of the theory, however, falls outside our current concerns.
37. It should be stated here that the model proposed in the previous section is not an abstract rationalization of the processes described by Changeux, but was first developed on independent theoretical grounds by Sinha (1984).
38. Rumelhart and McClelland (1986: 122ff.) maintain that PDP models are stated at the algorithmic level, in the terms of Marr (1982), but emphasize the interrelation rather than independence of computational and algorithmic levels; while also drawing attention to the 'emergent properties' intervening between implementational and computational levels. Multi-author citations, here and below, refer to chapters in Rumelhart *et al.* (1986).
39. Conventional computer memories are accessed by address-pointers, rather than content.
40. Although empirical findings are relevant to formulating epistemological questions, the history of science—and especially of cognitive science—demonstrates that they are rarely, if ever, decisive in determining answers.
41. For a wider discussion, not restricted to developmental issues, of PDP models and cognitive science, see ch. 4 of Rumelhart *et al.*, 1986.
42. Plasticity at this level would permit, for example, selective stabilization of visual–manual pathways in deaf children acquiring sign language.
43. Bates and MacWhinney (1987) have proposed a 'variation-competition' model of language acquisition which combines a functionalist perspective on grammar with a neo-connectionist learning mechanism. This model is based on principles entirely compatible with the theoretical position advocated here, but since it is still at an early stage of development I shall not discuss it in detail.
44. Although I emphasize that the theory I propose does not prescribe any specific linguistic theory, it should nonetheless be clear that it is more compatible with functional (Dirven and Fried, 1987) and cognitive (Langacker, 1987) grammars than with generative grammars.

5 Context: Background, Presupposition and Canonicality

Clay is moulded into vessels, and because of the space where nothing exists we are able to use them as vessels. Doors and windows are cut out in the walls of a house, and because they are empty spaces, we are able to use them. Therefore, on the one hand we have the benefit of existence, and on the other, we make use of non-existence.

Lao Tzu[1]

True, one portrait may hit the mark much nearer than another, but none can hit it with any very considerable degree of exactness. So there is no earthly way of finding out precisely what the whale really looks like.

Herman Melville[2]

Theoretical and practical reasoning

The problem explored in this chapter is that of the relationship between an object under representation, and its context, background or setting. The problem is not, of course, a new one. Traditionally, however, psychologists have investigated the relationship between object (figure) and field (ground) in a perceptual space viewed from outside by a non-participating observer. My concern, in contrast, is with the situation of an object within an intersubjective cognitive space constituted by the background[3] framing *both* subject and object at a given moment. In contrast to the Cartesian dualism of the observer/observed relation, I emphasize the inter-relatedness of subject and object, within a discursive framework informed by practical interest. And in contrast to the solitary ego of the observing I/Eye, I emphasize the collaborative and negotiative movement of co-operative reasoning in the establishment of shared meaning. My point of departure, then, is a communicative, rather than a reflectionist, epistemology; involving a shifting, discursively located subject, rather than a fixed and constitutive *Cogito*. From this perspective, what best characterizes 'background' is being *presupposed* in communication: to paraphrase Kant (see Chapter 1), although not all background knowledge emerges *from* communication, it is established *as* presupposed background *through* communication.

I also examine the developmental aspects of background knowledge, and its relationship to other aspects of conceptual development, particularly the acquisition of semantic knowledge. There is a long history in developmental psychology of seeing development as a process of progressive liberation of thought or representation from its context; involving movement away from 'spontaneous', context-bound, pre-logical thinking, towards systematized, generalized and 'scientific' conceptualization (see Chapter 3). Piaget, in his early work, associated syncretism, realism, animism and so forth with infantile egocentrism, but emphasized also that 'from the positive point of view [egocentrism] consists in the ego being absorbed in things and in the social group' (Piaget, 1959 [1926]: 271). In the work of Vygotsky, we find a related view of development, coupled with a psycho-pedagogical theory in which conceptual development is driven in part by the communicative and educative interventions of adults: 'the rudiments of systematization first enter the child's mind by way of his contact with scientific concepts and are then transferred to everyday concepts, changing their psychological structure from the top down' (Vygotsky , 1986: 172–173).

More recently, Donaldson (1978) has distinguished between thinking which is 'embedded' in 'human sense'—'dealing with people and things in the context of fairly immediate goals and intentions and familiar patterns of events' (p. 76); and 'disembedded' thinking unsupported by such human sense. Like other authors, she relates the ability to engage in disembedded thinking to literacy, positive social valuation and educational success (cf. also Olson, 1977; Scribner and Cole, 1978; Wells, 1981). Other authors have referred to 'embedded' thinking in terms of 'socio-dialogic sense' (Karmiloff-Smith, 1979) and 'social logic' (Sinha and Walkerdine, 1978). Nelson's (1985) account of semantic development proposes, similarly to Vygotsky, a progression from contextually derived event representations, to culturally imposed (and learned) systems of *sense-relations* (see Chapters 1 and 2). Without commitment to any specific initial formulation, I shall henceforth simply refer to these two modes, or poles, of thinking as *practical* and *theoretical* reasoning. Most accounts agree in:

1. According developmental priority to practical reasoning.

2. Emphasizing the social-interactive, communicative pragmatic aspect of practical reasoning.

3. Analysing practical reasoning in terms of *local, temporary and extended* representations of scenes, events or action sequences; rather than in terms of the *universal, stable and particulate* (though systematic) representations characterizing theoretical reasoning.

4. Within the extended representations characteristic of practical reasoning, representing conceptual and inferential linkages in terms of *contextual contiguity*, rather than causality or logical necessity (Bates *et al.*, 1979; Nelson, 1985; Schank and Abelson, 1977).

Is background knowledge, then, to be equated with practical reasoning? Or are there two types of background knowledge, practical and theoretical? And what developmental processes underlie the transition from practical to theoretical reasoning?

A cogent psychological analysis of the concept of background knowledge has been provided by Larsen (1985), who identifies and resolves a number of ambiguities and confoundings in previous treatments of this topic. He begins with the observation that background knowledge is best viewed not as a specific type of knowledge, but rather as:

> a *function* that knowledge may fulfill in relation to the subject's current activity ... background knowledge is recruited from a much wider repertoire of knowledge, our total *knowledge base* ... At any given moment, a particular subset of the knowledge base is actively used as background knowledge, but there is no permanent distinction between that subset and the remaining, inactive part of the knowledge base. Similarly, there is no permanent distinction between background knowledge, and what may be called foreground knowledge—that is, the knowledge and the information from the environment that is the focus of one's attention at the moment (p. 26).

Larsen goes on to suggest that although the background function may be served by any subset of the total knowledge base, there does exist a typology of background knowledge used by subjects in different contexts, distinguishable in terms of both the degree and the kind of *specificity* of the knowledge represented. He criticizes Tulving's (1972) distinction between 'episodic' and 'semantic' memory by pointing out that (a) is it unclear whether this distinction refers to distinct functional systems or to the type of

Personal specificity	Situational specificity	
	Situated	Desituated
Personal	Episodic, Autobiographic	Self, Identity
Depersonal-ized	Historical, Factual	Semantic, Conceptual

Reproduced by permission from Larsen (1985: 29).

Figure 5.1: Specific and general knowledge: the interaction of situational and personal specificity

knowledge stored in the memory mechanism(s); and, with respect to the latter, (b) Tulving confounds *personal* (autobiographic) with *situational* (indexical) specificity. Not all situationally specific knowledge, argues Larsen, is directly tied to personal experience. Equally, not all personal knowledge is indexed to specific episodes: knowledge of one's identity, dispositions and so forth is personal but 'desituated'. The orthogonal interaction between these two types of specificity, personal and situational, yields the fourfold classification of *types of background knowledge*, shown in Fig. 5.1.

Finally, Larsen makes a number of interesting remarks about knowledge development and acquisition, suggesting that the process of progressive decontextualization of cognition involves both 'desituating' and 'depersonalizing'; and that, while not all 'general' knowledge is acquired by generalization from directly experienced episodes (e.g. knowledge acquired through instruction), nonetheless 'episodic knowledge is basic in the sense that it delivers the basic materials for the development of more general knowledge; although it is not the one and only source of knowledge, it is precisely an indispensable background for all other forms of knowledge' (p. 30). If this is accepted, then a plausible developmental story begins with the type of knowledge characterized as autobiographic and episodic, in the upper left cell of Fig. 5.1, with representational development radiating outwards, via processes of 'desituating' and 'depersonalizing', to permit the elaboration of the other three types of knowledge.

Valuable as this analysis is, there remain two basic problems. The first is, quite simply, how this developmental process occurs. The second question is related to this: what is the nature of 'episodic' background knowledge—what internal structure, if any, does it possess? Episodes, after all, are not simples, but complexes of actions, objects, processes and events. An important issue, then, is to attempt to trace the developmental relationships between the relational knowledge permitting the encoding of event structures as holistic units, and the object knowledge permitting the analysis of episodes and scenes into stable components.

My main concern here will be with object knowledge, in the sense of classification: that is, with the establishment of categorial equivalence between objects, and the criteria for equivalence judgements.[4] Equivalence involves both similarity and difference, and in the next section, I examine the role of background knowledge in the selection of features of objects as *significant*: that is, as signifiers of difference and similarity. I do so with reference to a literary text: Chapter 32, 'Cetology', of Herman Melville's *Moby Dick*.

Background knowledge: inside or outside the whale?

In this text, Melville debates whether the whale should be classed as a fish or as a mammal. He notes that Linnaeus had declared in his *Systema Naturae* that whales were to be distinguished from fish 'on account of their warm bilocular heart, their lungs, their movable eyelids, their hollow ears, *penem intrantem feminam laclantem*'; and, ultimately, by recourse to natural and divine law. Melville, however, chooses to reject these arguments grounded upon the 'internal respects' of the whale, submitting rather to the judgement of Nantucket mariners, for whom the whale, being a creature of the sea, is properly to be classed as a fish. He goes on to define the whale 'by his obvious externals, so as conspicuously to label him for all time to come', as '*a spouting fish with a horizontal tail*'.

The Linnaean and Melvillean classifications share a common concern with structural description. The one, however, refers back, as Melville notes, to 'internal respects'—morphology, or structural relations *within* the object to be classified; whereas the other refers back to *external* relations—habitat or *context of situation* exterior to the object. Two different types of background knowledge are being invoked for the purposes of classification. To pursue the natural history metaphor, we might call one *endostructural*, referring to invariants, or at least lawful constraints, applying to the determinate form of the isolate or particulate object; and the other *exostructural*, referring to the behaviour of the object in relation to other objects, and especially in this case in relation to human needs, values, purposes, intentions and institutions.

Science and the practice of common sense

Science, as commonly understood, consists in the effort to *correlate* endostructure and exostructure descriptions, by explaining the behaviour and functioning of structures in terms of agreed criteria (physical laws, evolutionary adaptations, etc.). Such criteria are, in the usual view, supposed to exclude reference to human interest and value. The emphasis of science, on this account, is predominantly upon endostructure: a description is held to be a satisfactory explanation only in cases where the behaviour or functioning of the object (exostructural description) is seen to be lawfully grounded in, and governed by, the 'internal respects' of the object. Science proceeds by analysis; at least that has been, since Aristotle, the dominant view[5] and, at least in the case of the natural sciences, the dominant mode of scientific practice. Ever since Francis Bacon's strictures, in the *Novum Organon* (1960 [1620]), against the influences of the Passions and other 'Idols of the Tribe', and of Language and other 'Idols of the Marketplace'—influences which, in his view, could only be detrimental—the scientific method has also been conceived in opposition to human (subjective and social) value and intent (Larrain, 1979).

The human sciences, and to some extent the life sciences, differ from the physical sciences (at least as classically conceived), in that they place more emphasis upon exostructural description and the analysis of relations, than upon the analysis of essential cause. Linguistics is the paradigm example, insofar as since Saussure it has been a tenet of linguistic science that the identity of linguistic elements is wholly relational, that is, exostructural. The same emphasis upon exostructure underlies Marx's dictum that the human essence is nothing other than the ensemble of social relations. In the life sciences, a major aspects of the current debate on neo-Darwinism (see Ho and Saunders, 1984; and Chapter 4) centres upon the explanatory adequacy of accounts of evolution in which exostructural features (phenotypic behaviours) depend unidirectionally upon endostructure (genotypes), and in which context (natural selection) plays a merely limiting role. It might be suggested that, in general, reductionist theories tend to emphasize endostructure, and holistic and interactionist theories to emphasize exostructure. In the terms of Pepper's theory of 'root metaphors' (Pepper, 1942), a concern with exostructure and relations would seem to fit naturally with a 'contextualist' world view; and a concern with the analysis of the 'thing in itself', or endostructure, with a 'mechanist' world view. The former viewpoint will tend to emphasize the uniqueness, relativity and self-production of culture, and to question the notion of 'human nature'; the latter will tend to emphasize the grounding of behaviour in universal psychological, biological and (ultimately) physical processes.

There are, however, many species of 'reductionism', sociological and formalist as well as essentialist. Recently, in the social sciences, it is structuralist theories which have been criticized as mechanistic, and as excluding experience, value and subjectivity (Thompson, 1978). Generative Grammar, to take a non-sociological example, could also be viewed as a scientistic reformulation of the Saussurean project, in which structural relations themselves are derived from 'essential', formal properties such as predication and recursion (Moore and Carling, 1982). The single dimension of endostructure-exostructure seems, in any case, inadequate to accommodate Pepper's other 'root metaphors', 'formism' (which approximates to empiricism), and 'organicism', with its integrationist and developmental implications.[6] Nevertheless, bearing such problems in mind, the dimension, or polarity, of endostructure *versus* exostructure can serve as an important organizing principle for the analysis of background knowledge, representation and signification.

To return to this analysis: everyday practical reasoning, too, seeks to correlate endo- and exostructural description; but from the opposite, or if you like, complementary, point of view to that adopted by science and theoretical reasoning. The point of view adopted by practical reasoning is that of momentary and historically located human interest, informed by needs, desires, motives and beliefs, and eventuating in intentions, which in turn occasion actions.[7] Suppose that I wish to put a shelf up on a wall: I must seek out something that has the form of an angled bracket, and the intrinsic strength to support the shelf and its intended receipts. Whereas scientific reasoning (*theoria*) takes as its criterion for the successful resolution of a problem the discovery of previously invisible or unknown endostructural properties, practical reasoning (*praxis*) resolves problems by setting up a representation of an ideal or end state, and realizing it as closely as is practicable in the circumstances. As Bruner (1971: 65) puts it: 'skilled activity is a programme specifying an objective or terminal state to be achieved, and requiring the serial ordering of a set of constituent, modular subroutines' (see also Schmidt, 1975; Sinha, 1982a).

In very many real instances of problem solving, of course, theory and practice combine and overlap. In the same way, although the focus of theoretical reasoning is usually on endostructure, the actual movement of reasoning in the exploration of the properties of the world involves a dialectic between exo-to-endo and endo-to-exo directional vectors. A previously unknown chemical compound may be discovered, isolated or synthesized. In order to infer its endostructure (molecular composition), the scientist must rely upon systematic or serendipitous tests to discover its exostructural (reactive) properties. Another, science fictional, example: an alien spacecraft leaves behind an exquisite but mysterious artefact. Scientists start by fiddling with knobs, then dismantle the object, hoping to discover 'what it is for' (exostructure) from 'how it works' (endostructure). Eventually, unable to fathom the complexities of the alien object (Is it a communication device? A navigational aid? A sophisticated environmental sensor?), one of the scientists takes it home and uses it as a paperweight. Actually, the object is the alien equivalent of a chess computer. Ignorant of the system of value and interest of an alien form of life, however, the terrestrial scientists are unable to formulate the appropriate hypotheses.

Compare with this the behaviour of young children when faced with a novel apparatus (Hutt, 1966), who typically first investigate its parts and their functional relationships (endostructure), then explore its potential to be assimilated into their own playful action schemes (exostructure). To use Piaget's terminology, adaptation of purposive action to endostructural properties of objects constitutes *accommodation*, whereas adaptation of objects to current purposes and their exostructural realization constitutes *assimilation*.

Piaget sees *signification* as arising essentially from accommodation, with assimilation serving to provide the 'signified' schematic meaning: 'Representation begins when there is simultaneous differentiation and co-ordination between "signifiers" and "signified". The first differentiations are provided by imitation and the mental image derived from it, both of which extend accommodation to external objects. The meanings of the symbols, on the other hand, come by way of assimilation, which is adapted representation' (Piaget, 1962: 3). This view of signification is entirely in accordance with Piaget's general theoretical approach, which understands representation solely from the point of view of the subject (and his/her own actions), rather than in terms of the *relationship* between subject and object. An alternative and more adequate view requires the analysis of signifying relations as they occur in *both* accommodatory *and* assimilatory moments of action and representation.

In the relation between subject and object, what from the subject's point of view consists of a conservative, motivated, assimilatory adaptation of the object to the subject's immediate goals, from the *object's* point of view consists of a relocation, or recontextualisation, of the object in a novel matrix of arbitrary exostructural relations. The object (as, for example, when an object is substituted for, or made to 'stand for' another) now *signifies this entire exostructural matrix*, to just the same extent that it signifies the imaginary object which is represents. Play, as well as imitation, involves signification, and the signified meaning resides in the exostructural matrix within which the object/signifier is embedded, and not merely in the mind of the individual subject. By the same token, what from the subject's point of view consists of the exploratory (thus arbitrary) accommodation of schemata to the properties of a novel object, from the object's point of view consists of a conformization, or bringing into congruence, of exostructural relations with the endostructure properties which naturally or *canonically* afford those relations. Thus, cups are meant for drinking, boots for walking, and so on; and their endostructural descriptions *represent*, as I argued in Chapter 3, their canonical exostructural relations (or use values).

As the last two examples illustrate, it is important to distinguish between natural kinds and artefacts as types of objects. In the former case, the motivation for the relationship between endo- and exostructures is provided by natural law. In the latter case, it is provided by social practice, informed by interest and value. Canonical rules, in general, are what govern relations between endostructure and exostructure in non-natural kinds, and their motivation is to be sought in the way in which human interest and value is both shaped by, and shapes, a natural environment. Cups, for example, do not occur naturally, but must be produced; yet there are certain physical constraints (natural laws) governing the range of possible endostructural realizations or representations of the exostructural or functional purposes, needs, values, intentions, rituals and so on which are designated by 'drinking'. Basically, a cup, to be a drinking vessel, must be a container; and containers, given the natural world which we inhabit, and which partly shapes our form of life, must possess a certain (endo)structure. In this sense we may refer to the cup as a canonical container. Canonical rules, in a fundamental sense, *mediate* nature and culture; they are thus epistemologically privileged grounds of representational and signifying activity. In Chapter 3, I argued that canonical rules also constitute (both phylogenetically and ontogenetically) privileged sites for the emergence of human representational abilities. Later in this chapter, I offer experimental evidence for this proposal.

Metaphor and metonymy

I have moved from the consideration of objects as simples, to the consideration of objects as (simultaneously) signs. It is, then, appropriate to note the correspondence between the analysis above of objects and their relational structures, and the analysis of the structure of the linguistic code proposed by Jakobson—an analysis which itself consists of an elaboration of the distinction drawn by Saussure between the paradigmatic, synchronic, and the syntagmatic, diachronic dimensions of language. According to Jakobson (1956: 60–61):

> Any linguistic sign involves two modes of arrangment.
>
> 1. Combination. Any sign is made up of constituent signs and/or occurs only in combination with other signs. This means that any linguistic unit at one and the same time serves as a context for simpler units and/or finds its own context in a more complex linguistic unit. Hence any grouping of linguistic units binds them

into a superior unit: combination and contexture are two faces of the same operation.

2. Selection. A selection between alternatives implies the possibility of substituting one for the other, equivalent to the former in one respect and different from it in another. Actually, selection and substitution are two faces of the same operation ...

The constituents of a context are in a status of continguity, while in a substitution set signs are linked by various degrees of similarity which fluctuate between the equivalence of synonyms and the common core of antonyms.

These two operations provide each linguistic sign with two sets of interpretants, to utilise the effective concept introduced by Charles Sanders Peirce: there are two references which serve to interpet the sign—one to the code, and the other to the context, whether coded or free; and in each of these ways the sign is related to another set of linguistic signs, through an alternation in the former case and through an alignment in the latter. A given significative unit may be replaced by other, more explicit signs of the same code, whereby its general meaning is revealed, while its contextual meaning is determined by its connection with other signs within the same message. The constituents of any message are necessarily linked with the code by an internal relation and with the message by an external relation.

The aspect of the linguistic code governing relations of *selection, substitution and similarity* between linguistic signs was designated by Jakobson as its *metaphoric* pole, indicating that metaphoric figures depend for their interpretation upon these relations. The aspect of the code governing relations of *combination, contexture and contiguity* was designated *metonymic*, indicating that the device of metonymy relies upon part–whole relations in an ordered, extended context.[8] Jakobson's studies, which ranged from aphasiological investigations to the study of figures and tropes in verbal and visual art, led him to conclude that not only language, but all sign systems, are organized in terms of metaphor and metonymy, and that stylistic and discursive devices and genres derive from the differential positioning and emphasis of messages according to these two poles.

How does Melville's text relate to Jakobson's analytic framework? It is clear that the background against which Melville invites us to view the whale is *metonymic*, emphasizing context and contiguity; and that the contrasting Linnaean framework is *metaphoric* in emphasis, depending for its classification of the whale upon the selection of morphological features which are similar to, and intersubstitutible with, those of other mammals. In general, then: the pole of endostructure/morphology in objects maps onto that of metaphor in signs; and the pole of exostructure/relational context in objects maps onto that of metonymy in signs. However, because both objects and signs may be analysed at different structural and discursive levels, the representation of the object/sign along the polarity metaphor-metonymy is best seen as a function of the *relationship* between sign/object and background. This can be demonstrated by way of a further analysis of Melville's text.

Worlds of difference: from phenomenon to proposition

In proposing his definition of the whale as '*a spouting fish with a horizontal tail*', Melville makes a series of qualifying remarks about the definition, as follows:

1. The definition characterizes the whale 'by his obvious externals';
2. It characterizes the whale 'so as conspicuously to label him';
3. It characterizes the whale 'for all time to come'.

I shall discuss these remarks in turn.

1. The characterization by 'obvious externals' suggests at first that Melville is referring directly to exostructure: but, while we might stretch the point and concede that *spouting* is a behavioural-functional feature, *horizontal tail* is quite clearly a morphological one. And what are we to make of 'fish'? Melville *knows* that the whale is not a fish, in a strict sense. It seems rather that he is appealing to our *presuppositions* regarding fish, their habitat, nutritional value for human beings, typical shape, etc. The point is not that Melville rejects endostructural ('featural') criteria *per se*, but that he rejects the Linnaean ('scientific') approach to them, namely that the (endo)structure of 'internal respects' is ultimately determinant of the correct classification of the objects bearing these features. The most appropriate formulation might be as follows. Melville intends by 'obvious externals' to mean the *phenomenal form* of the whale, and his characterization of phenomenal form falls under the aspect of presuppositions associated with exostructural relations—and, in particular, exostructural relations from the point of view of practical human interest.

Phenomenal form, then—the way that objects *present* themselves to our *re-presentation*—is neither endostructure nor exostructure. It is a meeting ground, or interface, of a particular and special kind, which affords us a view of the object under both its endostructural and exostructural aspects. Melville's use of the presupposition-laden term 'fish'; his selection of the 'distinguishers' (Katz and Fodor, 1963) *spouting* and *horizontal tail*, implying the viewpoint and human interest of the masthead lookout of a whaler; and the whole tenor of his discourse serve to key in an exostructural background to the representation of the phenomenal form of the whale, motivated by historically and culturally located value and interest.

Phenomenal form can be likened to a visual illusion—such as the famous 'dog/rabbit' or 'old woman/young woman' figures. It confronts us, selectively, with the outcome of a historically structured dialectic, between the stability and resistance of material forms, and their malleability and utility in terms of human value and interest. Although the selection of a structural description for phenomenal form may tend towards either the pole of endostructural 'disembeddedness' or exostructural 'contextualization', this representational dialectic is constrained, in the case of natural kinds, by the fact that the *set* of features available for selection is finite and stable. This stabilizing and constraining function is fulfilled, in the case of artefacts, by canonical rules. It is likely, too, that for both natural kinds and artefacts, as the 'basic level' of categorization defined by Rosch (1977), the range of variation of representations of phenomenal form is restricted as a result of the evolution in the human species of mechanisms for the perceptual registration of 'affordances' in the Gibsonian sense. The fact, however, that discursive representations of phenomenal form are restricted by such mechanisms does not mean, however, that they are uniquely determined by them.

2. The characterization of the whale 'so as uniquely to label him' means that the labelling, or signification, should make the whale *stand out* from its background or context. Again, Melville's choice of terms illustrates his intention. This is to *select* and *distinguish* the whale from the host of creatures of the sea with which it (he) might be confused, from a particular viewpoint. This viewpoint, as we have seen, is that of the whaleman, not the scientific taxonomist. The architecture of the whale is described by Melville with regard to a contrast set comprising these organisms—fish—which might, by virtue of contextual, metonymic contiguity, be mistaken for the whale by the unwary or inexperienced mariner.

Melville does not *invent* the fluke or the spout of the whale; indeed, his imagined Linnaean adversary might well use exactly the same, objective features of whale morphology to guide apprentice scientists in the art of whale-recognition. In order to make the whale *stand out* from its metonymic context, Melville engages in a *metaphoric shift* in order to topicalize and focus certain features of endostructure. The combination of these features then constitutes a structural description adequate to the recognitory identification—or judgement of classification—of phenomenal form in context.

Melville is concerned primarily with *difference*, within a metonymically grounded discourse. The endostructure-oriented scientist, on the other hand, is concerned with a deeper structural *similarity*, permitting the placement of the whale within the class *mammalia*.[9] Melville rejects the 'deep structure' of Natural Order in favour of the law and lore of the whaleman. His own choice of 'law and order' is reflected in his selection of a gendered personal pronoun, designating the whale as a Significant Other in a primordial struggle between Culture—the masthead lookout—and Nature—the vast and undifferentiate mass of the sea and its denizens. Melville's text reproduces this law at a second order of removal. He forces the reader to confront difference by *introducing* a difference: between the 'cool' context of reflective, abstract understanding (the scientist in his study; charts, designs, specimens; the tabulation of natural forms reflecting the hidden purpose of the Divine Artificer); and the 'hot' pursuit of practical gain and value; order as a for-itself rather than as a pre-established in-itself; historical, practical reason, rather than timeless, transcendental Reason.

3. The characterization of the whale 'for all time to come' seems clear in its meaning. Melville is indicating that *all* whales, at *all* times and places, spout and have horizontal tails. He is, in effect, asserting the naturalness of the kind 'whale', and claiming the status of 'projectible predicates' (Fodor and Pylyshyn, 1981) for the spouts and flukes of whales. This claim marks a further discourse strategy, implicating not the foregrounded topic, but the presupposed background. Having established his description on practical-historical grounds, Melville now attempts to resituate this background, and to transform its status to that of timeless, theoretical (semantic) knowledge.

This discourse strategy exemplifies the point made by Larsen

(1985), that presupposed background is not defined by content. Rather, it is defined by the *function* of the content in relation to the topicalized foreground; by the agreed permissible procedures for its employment in reasoning and inference; and by the criteria which may be appealed to in the determination of the informational relevance (Grice, 1975) of particular messages. For example, the proposition '*That fish does not suckle its young*' is informative if it presupposes a Melvillean frame of reference, but not if it presupposes a Linnaean one. The Linnaean framework claims universality—it discursively forecloses the *possibility* of a mammalian fish. Melville's framework, on the other hand, is particular and context-bound; the *sense* of the proposition above is *local* to the Melvillean universe of discourse. Melville's discourse strategy—the claim that his definition of the whale holds good for all time—is therefore to be seen as a bid for universality. In Katz and Fodor's (1963) terms, Melville attempts thereby to shift the status of the features *spouting* and *horizontal tail* from that of distinguishers, to that of *markers*, of the family *whale* within the (Melvillean) order *fish*.

The semiotics of knowledge representation

Melville's discourse strategy is rhetorically effective, but epistemologically suspect. The features that he identifies are not properly to be identified as 'projectible predicates' in virtue of which whales 'satisfy laws', in the scientific sense. In point of fact, their very phenomenological salience gives good grounds for rejecting them as such, if we take science to be a movement 'beyond appearance'. Although features like *spouting* and *horizontal tail* can systematically be related to deeper, endo-structural properties of whales, designated by 'true' projectible predicates, Melville offers no systematic account of how this might be done. From a scientific, theoretical point of view, the flukes and spouts of whales constitute precisely part of the set of natural phenomena whose explanation is the task of science. They are not in themselves terms of explanation.

Nevertheless, given that whales *are* natural kinds, and that certain features of their phenomenal form *are* systematically related to endostructural properties designated by projectible predicates, and are *uniquely* so for the natural kind *Cetacea*, Melville's definition can properly be considered universal, in a non-explanatory, *practical* sense. What Melville offers may not be a scientific description of the whale, but it certainly is not an arbitrary one. Melville's description amounts to a set of generally applicable *recognitory criteria* for the determination of the extension class of the natural kind *whale*, whose intensional specification is informed by practical, exostructural, metonymic relations.

The most natural way of characterizing Melville's definition of the whale is as a *prototype* (Rosch, 1977). A prototype is a structural description in terms of phenomenal form. Although the background against which the whale is viewed by Melville is metonymic, the requirement that the features represented in the prototype should have cross-situational validity necessitates that *only* those (phenomenologically accessible) endostructural features be selected, whose combination is, precisely, uniquely (or maximally) typical of that object. In this way, metonymic combination and metaphoric selection interact in the construction of foreground as well as background. The metaphoric selection of the constituent features of the prototype is grounded in the metonymic context, and their metonymic recombination[10] *as* a prototype is dependent on the metaphoric 'lifting' of the signifying features out of their original (metonymic) context, to 'stand for' all (and only) those objects falling within an extension class defined by true (metaphoric) similarity.

On this account (which follows Rosch's closely), a prototype is a practical, or pre-theoretical, representation of a class in terms of a characteristic (ideally unique) *combination* of endostructural features, selected so as to be both phenomenologically accessible, and maximally informative, given a particular presupposed (background) metonymic context: i.e that context which typically frames the subject's encounters with the class. If this analysis is correct, prototypes (as mental representations) are not simple 'copies', but semiotic constructions possessing both an explicit and an implicit structure. Prototypic representations, on this account, should be seen as *cognitive signs* for recognitory schemata, constructed through the complementary processes of combination and selection.

Obviously, in the case of organisms such as whales, the selection and combination of prototype features is best effected in terms of what Melville calls the 'obvious externals' of the object. Other natural kinds may not solve the problem for human classification so easily. To recognize, even pre-theoretically, certain minerals for the purpose of smelting and working, or plants for medicinal and nutritional purposes, may require that practical reasoning refer as explicitly to metonymic, contiguity relations as to metaphoric, similarity relations. In all cases, however, metaphor and metonymy, as signifying relations, operate as complementary polarities in the representation of class membership; metaphoric focussing referring 'back' to metonymic context, and metonymic background permitting and motivating metaphoric selection. The 'core' of the prototype description, however, is the metaphorically selected and metonymically recombined 'topic', characterized by endostructural regularity, whether of 'obvious externals' or of less-obvious 'internal respects'.

Whereas, in the case of natural kinds, the endostructure–exostructure relation is motivated by 'natural laws' (e.g. evolution by natural selection) which apply independently of human interest and value (or at least have done so in the past), this is not the case for cultural products such as human artefacts. In that case, the endostructure–exostructure relation is motivated by canonical rules. And if, in the last instance, the determinant aspect of knowledge of natural kinds is their *independence* (at the level of endostructure) of human value and interest, the determinant aspect (in the last instance) of knowledge of cultural products consists in the exostructural canons and codes governing the functioning of these objects within the social order of purpose, value and meaning.

The preceding analysis now permits the further specification of the relations between the two distinctions which have motivated the analysis: on the one hand, between *practical* and *theoretical* reasoning, and on the other hand between *exostructure* and *endostructure*. As I have suggested, most analyses of 'embedded' thought tend to confound these dimensions, which are, in principle, independent. It is true that, as I have indicated, natural science tends to focus on endostructure; but, as has also been made clear, this conjunction does not hold for all modes of theoretical reasoning. In particular, those theoretical modes which are principally concerned with *formalization* may 'abstract' from the endostructural-metaphoric topic to the point where their 'objects' become nothing more than symbolic values over which algorithmic rules operate.

CONTEXTUALLY SITUATED	EXOSTRUCTURE METONYMY		ABSTRACT
	Sensorimotor Episodic	Algorithmic Formal	
PRACTICE			THEORY
	Iconic Prototypic	Semantic Propositional	
DECONTEXTUALIZED	METAPHOR ENDOSTRUCTURE		CONCRETE

Figure 5.2: Types of knowledge representation

By the same token, as I have also suggested, practical reasoning makes use of prototypic, metaphoric representations of specific object-classes, as well as of exostructural, episodic or functional representations of their metonymic contexts. To repeat an earlier point, both theoretical and practical reasoning seek to *correlate* exostructure and endostructure, rather than, in either case, concentrating on the one to the total exclusion of the other. If this is so, we can view the practical–theoretical and endostructure–exostructure dimensions as orthogonal, their interactions yielding the 2 × 2 tabulation of *types of knowledge representation* shown in Figure 5.2.

Fig. 5.2 does not map directly onto Fig. 5.1, since it conflates Larsen's (1985) important distinction between personal and situational specificity. Further, it is not intended as a characterization only of *background* knowledge, but as a semiotic typology of knowledge representation in general. The foreground–background distinction does, however, enter into the typology in several ways. First, those 'pre-propositional stances' (Chapter 2) which underlie more explict, intentional and propositional attitudes may be viewed as 'virtual' representations[11] informed by the polarity of practice: in this general sense, foreground in the sense of 'explicitness' maps onto theory, and background in the sense of 'implicitness' maps onto practice.

There is also, however, a more dynamic, discourse-constructive aspect to foreground–background articulation. In normal discourse, the *topic* is foregrounded, and linked by 'inferential bridges' (Clark and Haviland, 1977; Nunberg, 1979) to presupposed background.

In this respect, the representation (sign) in its metaphoric-endostructural aspect constitutes the topic governing the *selection* of background, and the presupposed, metonymic-exostructural context provides the elements available for *combination* through bridging inference. Thus, the metaphoric pole of signification corresponds (in general) to the topicalizing, rhematic, propositionalizing moment of discursive reasoning; while the metonymic pole of signification corresponds (in general) to the presuppositional, thematic moment of discursive reasoning.[12]

If this correspondence may be assumed to represent a 'natural' (or at least canonical) *Gestalt* structure for reasoning, a kind of psychological non-naturalness may be hypothesized to account for the cognitive difficulty involved in, on the one hand, 'presupposing' strictly semantic, non-contextually located knowledge; and, on the other hand, 'foregrounding' or topicalizing algorithmic procedures independently of metaphorically chained referents. Both of these cognitively difficult processes—which we might designate, respectively, as 'reasoning out of context' and 'reasoning in the abstract code'—involve articulations of metaphoric and metonymic polarities which violate the 'natural' metaphor-topic and metonymy-presupposition correspondences; and each is characteristic of different modalities of the employment of what Vygotsky terms 'scientific concepts'.

This hypothesis accords both with the many studies (some of which are cited above) indicating that a 'textual', 'semantic', 'code-oriented' cognitive mode is more difficult for children than a contextually situated, 'embedded' cognitive mode; and with the suggestion that one reason for the particular difficulty experienced by many children in learning the mathematical code is its uniquely *metonymic* character (Rotman, 1978; Walkerdine and Corran, 1979; Walkerdine, 1988).

The above analysis carries the further implication that—as was mentioned above—prototypic representations (in particular), and iconic representations (in general) constitute a semiotically more complex mode of knowledge representation than sensorimotor and episodic (event) knowledge, inasmuch as they implicate processes of selection and combination which 'lift' constituent elements out of their context of background. The developmental issues here are extremely complex, but the analysis is such as to suggest that the referential objects upon which language, as a symbolic system, is articulated, are not perceptual simples but cognitive signs—or, to put it another way, that perceptual structures such as prototypes are *cognitively constructed* and *semiotically mediated*.[13] Although I have concentrated thus far upon the semiotic processes underlying the construction of prototypic representations, it should be clear that the argument applies with equal, if not greater, force, to cognitive processes underlying the *literal* construction of variable exostructural relations from given, endostructurally typed objects (whether these objects are construed as signs or as physical objects).[14] If the arguments which I have advanced are correct, then theories which propose a non-mediated mapping between language and perception are fundamentally mistaken, whether from the point of view of mature psycholinguistic processing or from the point of view of language acquisition.

It is to language acquisition processes, and specifically to the acquisition of word meaning, that I now turn, in order to exemplify the analysis which I have provided in empirical studies.

Understanding utterances in context: the acquisition of spatial prepositions

In this section, I investigate the role of presupposed background (including canonical rules) in representational development. The analysis will be developed with reference to studies of children's comprehension of the locative prepositions *in* and *on*.[15] In particular, I shall examine the competing claims of the 'semantic features hypothesis', as modified in the 'partial semantics hypothesis' (Clark, 1973); and the 'functional core concept' hypothesis (Nelson, 1974). I shall suggest that both empirical and theoretical considerations support an account which stresses the role of canonical rules as a fundamental basis of conceptual and linguistic representation. Insofar as these rules are materially inscribed in form–function correlations of real objects, and socially inscribed in human interactions involving objects, they perform a crucial signifying role in the earliest stages of the integration of language, action and perception. I shall suggest that such an account demands a revision of the notion of 'context' as a 'variable' in opposition to internal, mental representational. In the present account, context is seen, not as a static property of the external environment, but as the dynamic product of organism–environment interaction. This dialectical model of cognition and context will also highlight the inherently social and communicative nature both of linguistic and cognitive development, and of the knowledge acquired in development.[16]

In her 1973 study of locative comprehension, Eve Clark required children to act out instructions containing the prepositions *in, on* and *under*, whereby toy animals were to be placed at various targets: box, tunnel, bridge, truck, crib, table, block, glass. Her data were consistent with the hypothesis that children's errors were governed by the application of an ordered pair of 'non-linguistic rules' (nlrs) which can be precisely paraphased:

Rule 1. If the target is a container, place the toy inside it.

Rule 2. If the target has a horizontal surface, place the toy on it.

Clark suggested that these rules were strictly ordered 'in the sense that Rule 1 is always applied. In the event that it fails, Rule 2 is applied next' (p. 168). The nlrs were conceived as being 'based upon the child's knowledge of the usual or expected spatial relationship' (p. 168) holding between given objects, and the ordering of the rules was seen as determining *both* comprehension strategies *and* acquisition order for the prepositions.

There is, however, a symptomatic ambiguity in the formulation of Rule 1: the term 'container' may refer either to the perceptual *appearance* of the target—that is, whether or not it presents a visible cavity suitable for containing another object; or it may refer to the canonical *function* of the target, as an object customarily used as a container, with suitable (visible or invisible) design features. In the former case, the orientation and immediate phenomenal appearance of the target object, will be relevant to whether or not Rule 1 is instantiated; in the latter case, it will be irrelevant.

Rule 2, however, is unambiguous: it is clear that reference is being made to a visible horizontal surface. In fact, Clark's discussion as a whole indicates that she intended both rules to refer to the (explicit) appearance of the array, and not to its (implicit) function.

In contrast to the partial semantics hypothesis, Nelson (1974) advanced the hypothesis that the core of the conceptual representation of objects consist in the specification of the dynamic, functional relationships into which they enter, assigning a separate role to their perceptual attributes as the basis for their *recognition* as instances of the functionally defined concept.

In relation to the discussion in the previous section, it is reasonable to suggest that Clark's partial semantics hypothesis, in emphasizing the primacy of perceptual properties of single, static objects as the basis of both the child's mental lexicon, and developmental errors made in lexical production (e.g. over-extension) and comprehension, is interpretable as an argument for the epistemological and developmental primacy (indeed, innateness) of perceptual discriminatory capacities geared to relatively invariant endostructural object properties. In this view, exo-structural-relational knowledge is secondary to, and derivative from, endostructural knowledge.

Nelson's functional core concept hypothesis, on the other hand, is interpretable as an argument for the primacy of episodic and sensori-motor representations of exostructural regularities in the usage and contexts-of-encounter of objects, in relation to other objects and to human agents (see Nelson, 1983, 1985).

Such a re-interpretation of these two accounts of semantic development is advantageous for a number of reasons. First, because there is no *a priori* reason for supposing that *both* endostructural and exostructural invariants and regularities are not implicated in early representational development. Second, because the grounds of the argument are shifted away from a simple opposition between 'form' and 'function', and between action and perception, towards an exploration of their interactive relations in the language-acquiring child. Third, because such a reinterpretation suggests that a strict, exclusive mapping between either perceptual, or functional-episodic representations, and semantic knowledge, is unlikely to hold either for children or adults.[17]

The dimension endostructure–exostructure is not the only one distinguishing the perceptual (P) and functional (F) hypotheses. These also include:

1. Static vs. dynamic representations of perceptual arrays. The P-hypothesis bases its account of both children's lexical acquisition strategies and their error strategies upon the supposition that children's hypotheses about word meanings relate to static array features. The F-hypothesis, on the other hand, proposes that children's hypotheses about word meaning relate primarily to the

dynamic features of the array (see also Greenfield, 1982).

2. Criterial-features vs. holistic-probabilistic representations of word meanings. The P-hypothesis, in postulating a direct mapping between perceptible array features and the semantic features constituting lexical meanings, presupposes the secondary hypothesis that semantic intensions can unambiguously and exhaustively define extensional range. The F-hypothesis, on the other hand, distinguishes between the functional-relational information that is definitive of the concept, and the perceptual information permitting the recognition of instances of the concept. The latter is conceived in probabilistic terms, similarly to the proposals of Rosch (1977) regarding prototypic representations.

A major problem in assessing the two hypotheses lies in the fact that the formulations of both contain supplementary hypotheses which render the two theoretical positions difficult to subject to decisive empirical test. Thus, for example, Nelson (1974) proposes that concepts are initially formed on the basis of functional relational information, but that new instances of the concept are identified on the basis of perceptual information. Equally, Clark (1973) suggests that children supplement their perceptually defined 'general' knowledge of locative relations with specific knowledge or particular objects' canonical orientations.

In relation to children's acquisition and development of spatial prepositions, there are, however, two fundamental points at issue. The first is the issue of the context specificity vs. the context independence of Clark's proposed nlrs. Since these rules were conceived by Clark as providing a basis for an overall order of acquisition of the prepositions *in, on* and *under*, as well as for children's errors of miscomprehension, there is no doubt that she intended them to apply independently of the specific objects composing the array.

Wilcox and Palermo (1975), however, suggested that the conformity of Clark's subjects to the nlrs was a contextually determined artefact of the particular objects she used in her experiment. In a study using different objects, they found no evidence that the nlrs accounted for all errors, and no evidence of an a-contextual ordering of difficulty of the three prepositions independently of the objects composing the array; a finding which is consistent with an unpublished study by Sinha and Walkerdine (1975). Wilcox and Palermo concluded that children's comprehension strategies were governed by two response tendencies: '(1) the tendency to make the simplest (and thus the easiest) motor response; and (2) the tendency to put the objects in their most normal (congruent) contextual relationship' (p. 251). Wilcox and Palermo's findings, then, are consistent with an account stressing the primacy of canonical relational knowledge, and by implication with the functional core concept hypothesis, but cannot be read as providing decisive evidence, since the motor complexity of required response patterns and canonical/non-canonical orientation of the target object were not controlled independently of the latter's functional and perceptual characteristics.

The second fundamental issue is the relation between perception, conceptual representation, and semantic representation. The P-hypothesis, which stands, as Nelson (1974) noted, in direct descent from the empiricist theories of Locke, proposes a direct and unmediated derivation of (semantic) knowledge from perception, basing conceptual and linguistic representation upon 'qualities' rooted in sense experiences. The incorporation of the nativist hypothesis that at least some features are innately specified is by no means incompatible with this perspective, as we have seen in discussions in previous chapters of the relationship between contemporary neo-rationalism and traditional empiricism. It is, of course, precisely this epistemological stance which was repeatedly criticised by Piaget as being unable to account for dialectical and structural, as opposed to merely incremental, cognitive transformations.

The F-hypothesis, on the other hand, as Nelson (1974) acknowledges, has extensive affinities with Piagetian theory. However, it also differs from Piagetian theory in that its concern is with the acquisition and development of specific conceptual knowledge, of particular objects and object classes, rather than generalized, schematic knowledge of the abstract 'object concept'. This aspect of the theory offers the possibility of overcoming the limitations of the methodological individualism of Piaget's theory, through the social and canonical specification of 'function'.

If we assume (correctly, I think, in the light of Nelson, 1985) that Nelson's F-hypothesis is referring to canonical rather than contingent functional relations, then we must face a further methodological problem: that is that, in the case both of natural and artificial kinds, behaviour or function and prototypic form tend to co-vary. On this basis, some authors have questioned whether Nelson's theory actually adds anything to prototype theory, since (a) Rosch (1977) cites involvement in similar motor programs as a characteristic of objects belonging to a 'basic level' category and (b) the co-variance of form and function in the real world renders the theory difficult to test empirically against an account stressing perceptual similarity (Bowerman, 1978; Barrett, 1982). A final, more substantive problem for the F-hypothesis concerns how the system of relational knowledge underlying, for example, the use of spatial prepositions, is developmentally differentiated from the 'functional core' of object representation.

In summary: (1) although the functional core concept account is compatible both with empirical evidence, and with the overall theoretical approach of this book, it is difficult to differentiate its empirical predictions from those of prototype theory; (2) although the account is based upon the primacy of relational knowledge, it does not specify how relational knowledge is differentiated from object knowledge, in such a way that it can permit novel functional relations to be constructed between objects. An empirical test of the functional hypothesis, then, ideally requires a design in which appearance, orientation, canonical function and complexity of required motor response are independently controlled. The experiments reported below are a first step towards meeting these requirements.

Experiment 1

Subjects: 137 children were tested, 70 boys and 67 girls, divided into five age groups.

Group 1, 36 children, mean age 18 months, range 16–20 months;

Group 2, 36 children, mean age 24 months, range 22–26 months;

Group 3, 31 children, mean age 30 months, range 28–32 months;

Group 4, 22 children, mean age 36 months, range 34–38 months;

Group 5, 12 children, mean age 45 months, range 42–51 months;

Apparatus: A cylindrical plastic beaker, blue (in half the experimental sessions) or orange in colour, measuring 10 cm high

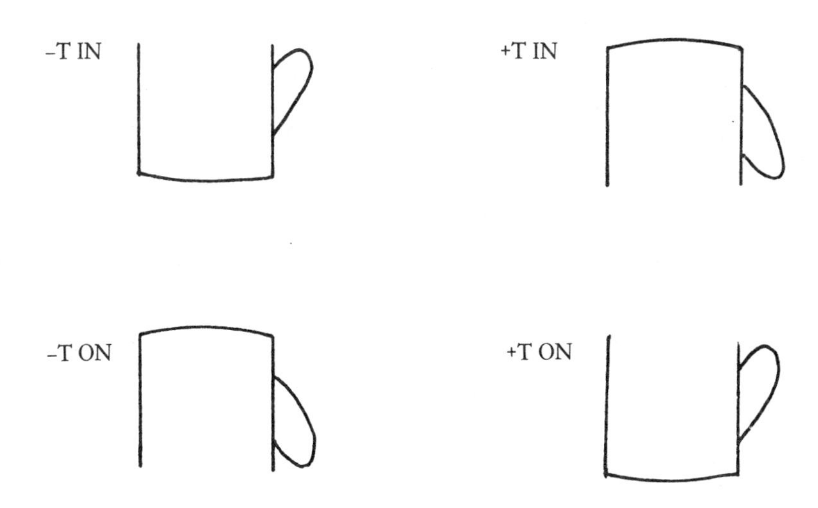

Figure 5.3: Trial types for experiment 1: initial configurations

by 28 cm circumference. A coloured wooden cube, 4 cm sides.

Procedure: The experimenter seated himself opposite the child, and placed the beaker, either upright or inverted, in front of the child within easy reach. The experimenter then handed the cube to the child, requesting him or her to 'put the brick in/on the cup' (NB *under* trials were also carried out for both Experiment 1 and Experiment 2, but these will not be discussed here: see Sinha, 1982a). For present purposes, for each subject there were four trials in all, two for each of the prepositions *in* and *on*, one of which required a transformation (+T) of the initial orientation of the cup, and one of which did not require a transformation (–T) of the initial orientation of the cup, in order to produce a correct response (Fig. 5.3).

Rationale: The experiment permits an initial test of the perceptual (P) and the functional (F) hypotheses in terms of whether children's error strategies are determined by the perceptual or the functional characteristics of the array. It also permits a test of the specific hypotheses advanced by Clark (1973), in the form of her proposed non-linguistic rules. The experiment also permits an analysis of the extent to which motoric complexity of required response interacts with other non-linguistic factors in determining task complexity.

Table 5.1: Comprehension of *in* and *on* using cups as targets, by age group

Age group	Group 1		Group 2		Group 3		Group 4		Group 5	
Mean age	1.6 yrs		2.0 yrs		2.6 yrs		3.0 yrs		3.9 yrs	
Group N	24		33		28		21		12	
Trial Type	–T	+T	–T	+T	–T	+T	–T	+T	–T	+T
Responses										
In correct, On correct	15	2	22	5	27	5	19	15	12	9
In correct, On wrong	9	12	9	13	1	8	1	1	0	0
In wrong, On correct	0	0	2	2	0	1	1	2	0	0
In wrong, On wrong	0	10	0	13	0	14	0	3	0	3

Results: The results are shown in Table 5.1. All errors involved either the child making an *in* response to an *on* instruction, or vice versa. The data are presented in the form of individual strategy outcomes. Data are presented only for those children completing all four trials.

Discussion: The main effect is that for the two youngest groups, *in* trials were easier than *on* trials for both –T and +T conditions. This effect is greater for the +T than for the –T trial types: (Group 1: $p = 0.002$, –T; $p < 0.001$, +T. Group 2: $p < 0.05$, –T; $p < 0.005$, +T). (Binomial test performed on error types where only 1 of the respective In and On trials was correctly responded to.)

For group 3, this effect was significant for +T trials ($p < 0.02$), but not for –T trials. There was no significant effect for either trial type, separately or combined, for Groups 4 and 5.

The *in*-superiority effect may be explained in terms of either of the two hypotheses.

The P-hypothesis proposes that: (1) *in* is acquired before *on*; (2) errors will be determined by the perceptual characteristics of the array through the application of the ordered nlrs. The P-hypothesis therefore predicts that:

1. For –T trials, both *in* and *on* instructions should be correctly responded to at all ages, regardless of whether or not the child understands the preposition, since in both cases, the application of the nlrs will yield a correct response. This was not in fact the case: this response pattern was only gradually acquired, increasing significantly in frequency with age ($\chi^2 = 17.75$; $p < 0.001$). Many of the younger children righted the cup in response to the *on* instruction, and produced an *in* response. This response pattern was significantly more frequent in the two youngest groups of children than in the three oldest groups ($\chi^2 = 15.4$; $p < 0.001$). This response tendency was also observed by Clark (1973), who accounted for it by proposing the supplementary hypothesis that children have knowledge of the canonical orientation of objects. I discuss this proposal below.

2. For +T trials, there will be a three-stage developmental pattern. Stage 1 subjects will understand neither preposition, and consequently the application of the nlrs will lead to errors for both *in* and *on* trials. This was in fact a frequent response strategy for the first three groups of children, accounting for 44 per cent of all responses. Stage 2 subjects will comprehend *in* but not *on*, and consequently will respond correctly to *in*, but wrongly to *on* instructions. This was the second most frequent response strategy in the first three groups, accounting for 39 per cent of all responses. Finally, Stage 3 children will respond correctly to both instructions. This correct pattern indeed increased regularly and significantly from 8 per cent of all responses at age 18 mo. to 75 per cent at age 45 mo. ($\chi^2 = 34.6$; $p < 0.001$).

The error patterns predicted for Stage 1 and Stage 2 by the P-hypothesis decreased abruptly at age 3.0, by which age performance was nearing ceiling, and the majority of children understood both prepositions for both trial types.

However, across the first three groups of subjects, i.e. those groups showing well below ceiling comprehension, neither Stage 1 nor Stage 2 response patterns showed any significant or even apparent tendency to decrease, casting doubt upon the status of the proposed nlrs as representing *developmental*, as opposed to merely contextual, strategies. The only response pattern which *did* decrease significantly over these three age groups was that in which children responded correctly to –T *in*, but incorrectly to –T *on* trials; which represented 37.5 per cent of all responses for Group 1, and 3.5 per cent of all responses for Group 3 ($\chi^2 = 9.32$; $p < 0.01$).

It was this error strategy which Clark suggested to be due to children supplementing the knowledge represented in the nlrs with knowledge of objects' canonical orientations. This explanation, however, is not wholly consistent with Clark's overall theoretical position, since the orientation of an object is specified by the position of particular endostructural *parts* of an object, in relationship to a background. Thus, it depends upon the endostructural asymmetry of a particular object (e.g. a cup has a cavity at the top, and a supporting surface at the bottom) which is precisely the crucial design feature of the object as a functional artefact: in this case, as a canonical container.

This particular error strategy, then, though it does not empirically contradict Clark's (1973) findings, is arguably more theoretically consistent with the F-hypothesis, expressed as a strategy of relying upon knowledge of the canonical *function* of the target object to supplement inadequate linguistic knowledge. By the same token, this strategy will also account for the same error pattern as that defining Stage 2 of the P-hypothesis predictions: namely, better performance for +T *in* than +T *on* trials. However, the F-hypothesis does *not* predict the other main error pattern found for the +T trials, in which both *in* and *on* instructions were responded to incorrectly.

One way to account for this pattern within the general framework of the F-hypothesis is to suggest that the tendency to respond to the canonical function of the target object (the canonicality effect) is modulated by the further tendency to make the easiest possible motor response, as suggested by Wilcox and Palermo (1975). If this is an independent effect, then one would expect that difficulty of motor execution should affect *in* and *on* trials equally within each age group, even though within each age group there should also be an overall superiority for comprehension of *in* over *on*. The two effects, the *canonicality effect* (c-effect), and the *motor difficulty effect* (m-effect), should then be additive, both also decreasing with age.

In order to analyse the interaction of the two effects, independent measures for them are required, for both –T and +T trials (c-effect), and for both *in* and *on* trials (m-effect). These measures can be expressed as the differences between either (c-effect) *in* and *on* correct responses (Dc), or (m-effect) –T and +T correct responses (Dm), proportionately to the total number of correct responses. Thus:

$$Dc = (IN) - (ON) / (IN) + (ON)$$

$$Dm = (-T) - (+T) / (-T) + (+T)$$

where bracketed symbols refer to the number of correct responses for the given trial type. For each age group, there will be two values of Dc, one for each of –T and +T trial types, and two values of Dm, one for each of *in* and *on* trial types, permitting analysis of the effect of each variable upon the other. The relevant raw data are shown in Table 5.2.

Table 5.2: Correct responses for each trial type, by age group

Age group/Trial type	–TIN	+TIN	–TON	+TON
1 N = 24	24	14	15	2
2 N = 33	31	18	24	7
3 N = 28	28	13	27	6
4 N = 21	20	16	20	17
5 N = 12	12	9	12	9

Table 5.3 displays values of Dc for both –T and +T trials, and Dm for both *in* and *on* trials, for each age group. Overall values of Dc and Dm are also shown for each age group.

Table 5.3: Canonicality and motor difficulty effects by age group and trial type

Age group	Dc–T	Dc+T	DmIN	DmON	DcALL	DmALL
Group 1	.231	.750	.263	.765	.382	.418
Group 2	.127	.440	.265	.548	.225	.375
Group 3	.018	.368	.365	.636	.108	.486
Group 4	0	–.030	.111	.054	–.014	.095
Group 5	0	0	.143	.143	0	.143

Clearly, there is a different developmental trend for the c-effect than for the m-effect. The former diminishes regularly in strength with increasing age, both overall and for –T and +T trials separately, reaching baseline at 3.0 years (Group 4). The latter also reaches baseline at this age, but does so abruptly; before that, it remains more or less of constant strength. Again, this is the case both overall and for IN and ON trials separately.

The response biases also clearly interact, in that the c-effect is more pronounced for +T trials than for –T trials, and the m-effect is more pronounced for *on* than *in* trials. These interactions can also be expressed numerically, as:

1. The dependency of the c-effect upon the m-effect. This will be expressed as the ratio Dc–T : Dc+T.
2. The dependency of the m-effect upon the c-effect. This will be expressed as the ratio DmIN : DmON.

In each case, a ratio approaching unity will indicate the independence of the first variable from the effects of the second, and a ratio approaching zero will indicate the strong dependence of the first variable upon the second variable. (NB A ratio exceeding unity would indicate an 'inverse' dependence, such that for example the m-effect is *less* pronounced for *on* than for *in* trials.) The respective figures for each age group are shown in Table 5.4.

The figures for Groups 4 and 5 cannot be very reliable, as they depend upon very small numbers. Nevertheless, the trends are clear, and different in the two cases. The c-effect's dependency upon the m-effect—that is, the extent to which response biases rooted in canonical function are affected by motoric task complexity—*increases* with age; whereas the m-effect's dependency upon the c-effect—the extent to which task motoric difficulty is affected by the canonicality/non-canonicality of the required end state—*decreases* with age.

Table 5.4: Dependency relations between c-effect and m-effect, by age group

Age group	1	2	3	4	5
Dc–T/Dc+T	0.308	0.289	0.049	0	–
DmIN/DmON	0.344	0.484	0.574	2.056	1.000

This analysis strongly suggest that the m-effect should be seen as a performance bias which becomes increasingly independent of the competence bias represented by the canonicality effect. This latter, competence-reflecting, bias is more likely to be evoked by motorically difficult than by motorically simple tasks, and is therefore only manifest in older children in tasks having a relatively high performance loading factor.

These findings support the suggestion by Wilcox and Palermo (1975), that Clark's (1973) proposed nlrs are artefacts of the interactions between (a) response biases rooted in the conceptual representation of objects in terms of canonical function; and (b) performance biases rooted in the difficulty or simplicity of the execution of particular motor programmes. The findings therefore tend to lend support to Nelson's (1974) functional core concept hypothesis.

Nevertheless, Experiment 1 cannot serve a decisive test, for a reason that I have already alluded to: the very coincidence between perceptual appearance, and functional specification, which is constitutive of the notion of canonicality, leads to rather similar predictions being made by the P- and F-hypotheses. In particular, since cups are canonical containers, a key prediction of the P-hypothesis—that *in* is mastered in comprehension, in this context, before *on*—is also predicted by the F-hypothesis.

Experiment 1 attempted to control canonical function and orientation independently of the motoric complexity of the action required of the subject. Experiment 2 goes one step further, by attempting to control the functional (exostructural) specification of the target object independently of its perceptual (endostructural) description.

Experiment 2

Subjects: The same five groups of children that took part in Experiment 1.

Apparatus: A set of 10 'Ambi' stacking/nesting cubes, the largest cube having sides of 7.5 cm, the smallest having sides 3 cm.

Procedure: This was a variant of that employed in Experiment 1. First, the experimenter sat opposite the child and introduced the apparatus as a set of toys, without naming them. Then he elicited the child's co-operation in playing either a *stacking* or a *nesting* game with the cubes. Every effort was made to ensure that, in the stacking game condition, the child did not gain experience of nesting, and vice versa. After playing for about five minutes, the experimenter chose a moment at which a full stack/nest had been constructed, and extracted the second largest cube from the configuration (Fig. 5.4). This was to be the target object, and it was placed in front of the child, within reach, either cavity up or cavity down, similarly to the cup in Experiment 1. The remaining cubes, still stacked/nested, were placed in the background, near the experimenter, and out of the child's reach. The experimenter extracted the smallest of the cubes from this array and handed it to the child, saying "Can you put this in/on there?" as in Experiment 1, with the trial order, as in that experiment, being randomized. After the child had completed all the trials in the first (Stack/Nest) condition, an interval of a few minutes elapsed, in which play in unrelated tasks took place, and the procedure was then repeated for the second condition. Order of administration of the two conditions was counterbalanced across subjects.

Experiment 2 was carried out on a separate occasion from Experiment 1, the order of the experiments also being counterbalanced across subjects.

Rationale: Unlike the cups used in Experiment 1, these cubes were inherently bifunctional, possessing no canonical function or

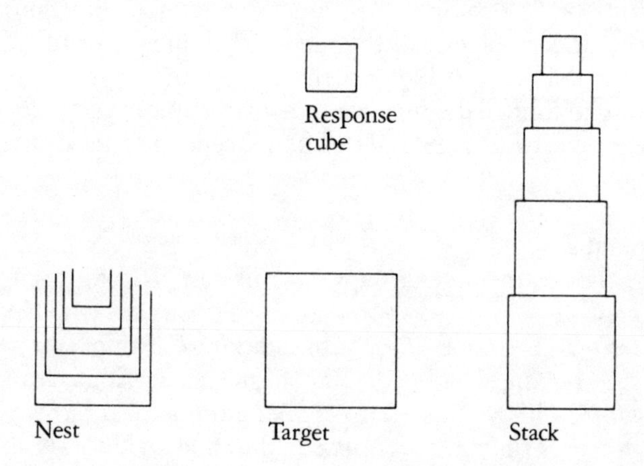

Figure 5.4: Target and Response (placement) cube in relation to constructed nest and stack arrays (Experiment 2: relative sizes)

orientation. In each of the two conditions, the child was presented with an identical array and identical instructions. The only difference between the conditions was the previous co-operative play activity which the child had experienced with the materials. Thus, this contextual/episodic *background context* must be responsible for any difference between response-biases in the two conditions. If such differences occur, this can be taken as evidence that the representation of objects in terms of their exostructural-functional relations can determine comprehension strategies independently of their endostructural-perceptual properties.

Results: I shall concentrate in this presentation upon those aspects of the data relevant to the hypothesis implied by the rationale, above: namely, that the response biases displayed by the children will differ significantly across the two conditions *Stacking* vs. *Nesting*. Tables 5.5 and 5.6 show the individual strategy outcomes for the two conditions, similarly to Table 5.1 for Experiment 1. As for that experiment, results for each condition of this experiment are only presented for those children completing all four trial types.

Discussion: Although there are some differences between the

Table 5.5: Comprehension of *in* and *on* using cubes as targets after playing a stacking game, by age group

Age group Mean age Group N	Group 1 1.6 yrs 13		Group 2 2.0 yrs 27		Group 3 2.6 yrs 24		Group 4 3.0 yrs 20		Group 5 3.9 yrs 12	
Trial type	–T	+T	–T	+T	–T	+T	–T	+T	–T	+T
Responses										
IN correct, ON correct	9	3	24	4	22	5	16	12	10	1
IN correct, ON wrong	3	3	1	6	0	5	1	3	0	0
IN wrong, ON correct	1	1	2	6	2	4	2	3	2	3
IN wrong, ON wrong	0	6	0	11	0	10	1	2	0	8

Table 5.6 Comprehension of *in* and *on* using cubes as targets after playing a nesting game, by age group

Age group Mean age Group N	Group 1 1.6 yrs 12		Group 2 2.0 yrs 26		Group 3 2.6 yrs 23		Group 4 3.0 yrs 20		Group 5 3.9 yrs 12	
Trial type	–T	+T	–T	+T	–T	+T	–T	+T	–T	+T
Responses										
IN correct, ON correct	10	1	23	1	22	4	20	10	10	2
IN correct, ON wrong	2	3	3	9	1	8	0	1	2	1
IN wrong, ON correct	0	0	0	3	0	2	0	5	0	0
IN wrong, ON wrong	0	8	0	13	0	9	0	4	0	9

response patterns displayed in Tables 5.5 and 5.6, there are also some common patterns. The most obvious is that the 'm-effect' noted for the cups experiment is also manifest in both conditions of Experiment 2, as manifested by the greater number of errors in both conditions for +T trials than for –T trials; and in the preponderance in the –T conditions of fully correct response patterns. This is the error pattern which would be predicted by the P-hypothesis, given the application of *both* nlrs.

However, since the cubes used as targets in this experiment were bifunctional (that is, lacking a canonical function), the P-hypothesis and the F-hypothesis make different predictions for the overall distribution, across the two (Stacking and Nesting) conditions, of error types. The P-hypothesis, in terms of the ordering of the 2 nlrs, will predict an overall *in*-bias, equivalent to the c-effect in Experiment 1, since the endostructural-perceptual description of the cubes is equivalent to that of the cups. The F-hypothesis, on the contrary, will predict that the episodically defined function of the object during the pre-task game will act as background knowledge leading to an *on*-bias for the Stacking

Table 5.7 Correct and incorrect responses to *in* and *on* trials for stacking and nesting conditions, by age group

	IN				ON			
	Stack		Nest		Stack		Nest	
	✓	×	✓	×	✓	×	✓	×
Gp. 1	18	8	16	8	14	12	11	13
Gp. 2	35	19	36	16	36	18	27	25
Gp. 3	32	16	35	11	33	15	28	18
Gp. 4	32	8	31	9	33	7	35	5
Gp. 5	11	13	15	9	16	8	12	12
Younger	53	27	52	24	50	30	38	38
% ✓	66		68		62.5		50	
Older	75	37	81	29	82	30	73	35
% ✓	66		81		73		68	

condition, and an *in*-bias for the Nesting condition. Table 5.7 shows the relevant results.

Table 5.7, as well as displaying data for each separate age group, also shows combined data. Groups 1 and 2 have been combined into a single (Younger) age group, (mean age 21 months), and Groups 3, 4 and 5 have been combined into a single (Older) age group (mean age 42 months). It is clear from the data that neither the P-hypothesis, nor the F-hypothesis, predictions are confirmed in any straightforward way. There is no overall *in*-bias, of the sort the P-hypothesis would predict; children at almost all ages give more correct than incorrect responses to both *in* and *on* instructions; and there is a tendency, as would be expected, for performance to improve with age. Neither does the experimental condition (Stacking vs. Nesting) appear to have any marked, global effect; the only sign of this is a depression of correct responses to *on* trials, for the Younger group, in the Nesting condition; and an elevation of correct responses to *in* trials, for the Older group, also in the Nesting condition. There are no significant within group differences between Stacking and Nesting conditions in the total number of correct responses to either *in* or *on* trials.

A more interesting picture is revealed, however, if subjects' error patterns are extracted from Tables 5.5 and 5.6, concentrating on those responses in which only one of the two instructions (*in* or *on*), for a given trial type, is correctly responded to, and the other one is incorrectly responded to. Combined data are displayed in Table 5.8.

Table 5.8: Error patterns by condition, trial type and combined age groups (Younger = Groups 1 and 2; Older = Groups 3, 4, 5)

Age Group	Younger				Older			
Condition	Stack N = 40		Nest N = 38		Stack N = 56		Nest N =55	
Trial Type	–T	+T	–T	+T	–T	+T	–T	+T
Error Type								
In ✓, On ×	4	9	5	12	1	8	3	10
In ×, On ✓	3	7	0	3	6	10	0	7

Both Younger and Older groups of children display an *in*-bias in their responses in the Nesting condition only; a bias which is stronger for the Younger group than for the Older group, and for the Younger group is most marked in the +T trials. In the Younger group this bias is abolished in the Stacking condition; in the Older group it is reversed in the Stacking condition to an *on*-bias. Table 5.9 recombines these data, across –T and +T conditions, for the two age groups and for the sample as a whole.

Table 5.9: Errors in stacking vs. nesting conditions

	Younger		Older		All S's	
	Stack	Nest	Stack	Nest	Stack	Nest
In ✓, On ×	13	17	9	13	22	40
In ×, On ✓	10	3	16	7	26	10

It is clear from these data that:

1. The *in*-bias predicted by the P-hypothesis is *only* manifest in the Nesting condition. In the Stacking condition, there is neither an *in*-bias nor an *on*-bias for the Younger group; and there is an *on*-bias, of approximately equal strength to the *in*-bias in the Nesting condition, for the Older group. The differences in error patterns between the Stacking and Nesting conditions approach statistical significance for each of the two groups individually ($p = .0435$, Younger; $p = .051$, Older), and are very clearly significant for the sample as a whole ($p < .001$) (Fisher test). This difference in response patterns to targets which are *identical* in perceptual appearance, can only be attributed to the differential *cognitive* representation and evaluation of the perceptual (endostructural) properties of the array according to the subject's prior experience of the object in different *exostructural* relational contexts of use. The demonstration of such a difference in young children's performance on a language comprehension task is powerful evidence in support of the F-hypothesis.

2. Although, as we have seen, it would be wrong to attribute subjects' response strategies solely to the perceptual appearance of the target object, this appearance does strongly affect responses. The most common error strategy on +T trials, for both Stacking and Nesting conditions, and for all age groups, is that in which *both in* and *on* instructions are incorrectly responded to: in other words, that response pattern in which children respond to the immediate perceptual (endostructural) properties of the target. Furthermore, insofar as the *in*-bias for the Nesting condition for the Younger group of subjects is only *abolished* in the Stacking condition, rather than (as for the Older group) being *reversed*, the experiment also provides some support for the ordering of the nlrs proposed by Clark.

It appears, then, that Clark's nlrs do represent a psychologically real aspect of children's processing of linguistic information in relation to a non-linguistic context; an aspect reflecting both an initial preference (at least for some object classes) for *in* over *on* relationships; and a reluctance to transform spatial arrays, perhaps because of the cognitive load thereby imposed. This manifests itself in a contextually determined, 'perceptual' strategy, relatively (not wholly) independent of a global preference for constructing one spatial relation over another. How might this strategy be explained? One suggestion might be to invoke Gibson's (1979) notion of affordance: a horizontal surface affords support, and a cavity affords containment. It is reasonable to suppose that evolution has equipped human infants with perceptual capacities tuned to such ecological affordances. It is, however, a long way from such perceptual mechanisms to conceptual representation, and even further to semantic representations.

Some kind of psychological representation of both objects and relationships as instances of *classes* is a requirement for conceptual development, and the way that classes are represented is not exclusively in terms of phenomenologically accessible endostructural features. It is also governed, as these experiments demonstrate, by the exostructural *background* against which objects are perceived and categorized. It is this background, whether represented as knowledge of canonical rules, or represented as 'local' rules (Donaldson, 1978) arising from specific interpersonal interactions, which determines, in interaction with other aspects of context, the *significance* of the perceptual properties of objects. In the early stages of development, the signification of specific perceptual features is largely determined by canonical rules relating function and form. Later, perceptual features are re-organized as cognitive signs functioning within plans and other intentional structures to permit the construction of non-canonical relations. Children's sensitivity to the establishment of context-sensitive local rules can be seen, in this respect, as a highly adapative cognitive mode—one which facilitates and underlies the process of appropriation of specific cultural practices and meanings.

As Nelson (1985) suggests, the acquisition of semantic knowledge proper induces further re-organizations of both the perceptual and the conceptual systems, permitting 'decontextualized' reasoning involving processes operating over symbolic objects, and the production of entirely new cultural forms. The particular psychological processes involved in these later developments fall outside my present scope, however; and I shall conclude by discussing the foregoing experiments in relation to the wider themes of this chapter, and the book as a whole.

Context, representation and social relations

Children's understanding of an utterance in relation to a context is governed both by current linguistic knowledge, and by inferences and hypotheses based upon the evaluation of the context. In this respect, the reasoning behind Clark's (1973) partial semantics hypothesis is supported by the experiments I have reported. However, these contextual inferences are not constructed only in relation to the isolated, particulate object which is the topic of the discourse. They are also constructed in relation to the *background* against which the child evaluates the *significance* of the topic, as an object (or representation) potentiating or inviting certain actions, interactions and relations. In other words, context is not a category external to, and separate from, the cognitive processes of the subject, but is jointly constituted by the presuppositions brought by the subject to the situation, and the elements of the attentional field which 'activate' the background presuppositions.

In cases where background knowledge is minimal—where the child has no experience of dealing with a particular object type—a developmentally productive strategy may be to rely upon the affordances provided by the endostructural features of the object, and to accommodate action to these affordances. This would be the case in the Hutt (1966) novel object paradigm, and would account, in conjunction with considerations of the motoric difficulty of particular actions, for the sort of strategy Clark (1973) formalizes as non-linguistic rules based upon perceptual information. In the usual case, however, background knowledge will specify also the exostructural (functional) relations into which the object (canonically or locally) enters. Such background knowledge is a decisive factor in the formation of the *significant context* (Sinha and Carabine, 1981) for the comprehension of the utterance.

In cases where immediate endostructural affordance and canonical exostructural relations coincide, for a given object in a given orientation, comprehension will be *canalized* towards the correct response. Such canalizations, which may be seen quite literally as privileged epigenetic pathways (Waddington, 1975), are provided by canonical rules interrelating endostructural and exostructural object profiles, which thus provide basic cognitive specifications of the phenomenal world.[18]

It would be a mistake, however, to see these rules as being

exclusively either 'in' the object or 'in' the mind of the individual subject. Canonical rules structure the world of the child by being *part* of that world—the *Umwelt* (Uexküll, 1957), or 'implicate order' (Bohm, 1980; Shotter and Newson, 1982), constituting the child's ecological niche, or environment for effective action. The structure of the *Umwelt* (or life space: cf. also Lewin, 1946; Palermo, 1982) is underdetermined by the laws of the non-historical, physical universe, for that world has itself been shaped and transformed by human history and culture. The *Umwelt* is saturated by mind—not just the mind of the individual child, but the mind of successive generations of human agents. Canonical rules *represent* historical consciousness and cultural order, canalizing the child's intellectual and communicative development towards the modes of subjectivity appropriate for the bearer of a particular, historically located culture.

Canonical rules are also *represented*: in artefacts, in utterances (including their syntactic as well as semantic structures: Slobin, 1981) and in structures of interaction—episodes of social exchange. The material world is not constituted, as Piaget can be read as suggesting, only by objects in the guise of Newtonian particles entering into logico-mathematical relations. It is also, and more importantly, constituted by objects (especially artefacts) entering into, supporting and constraining, *social relations*. Knowledge of the material world, then, is fundamentally *social* knowledge, and the process of acquiring that knowledge is also fundamentally social.

If canonical rules supply a stable presuppositional structure and background to the child's actions within the *Umwelt*, the *means* of learning and acquiring knowledge of these rules is provided by the child's participation in interactions with an adult partner, who can 'show' the world to the child by shaping it according to the rules: 'scaffolding' the interaction, to use Bruner's (1975) felicitous phrase, so as to draw out and render explicit ('relevate': Bohm, 1980) its implicit (implicate) order. In this way, the objects (in the sense both of 'things' and of 'goals') of the interaction are invested with significance. They become, for the child, material representations and signifiers of the rules, norms, values, rituals, needs and goals of the entire exostructural matrix within which, as significant objects, they are located. In short, they become part of a meaningful system of signs.

Coda

We are accustomed to think of 'things' as being material, and rules, relations and meanings as having a different, 'mental' status from the material 'things'. I have argued, throughout this book, that this Cartesian view is seriously misleading. We should rather see representation as *constitutive* of the (our) material world, than as a secondary structure supervening upon it. Only by so doing can we break with the dualistic assumption—deeply embedded in our current culture (including the subcultures of psychology and psycholinguistics)—that the mind of the individual can be studied independently of the social relations constituting historically located subjectivities.

This dualism underlies the chimerical goal which beckons the cognitive sciences: to delineate, once and for all, the Truth of Representation, stripped bare of the 'contingencies' of history. The Representation myth, as we might call it, is projected not only upon the Subject, but also upon the Scientist, as the ultimate arbiter of the subject's powers. The Representation myth, in denying the historical, interpretive moment of psychology as an irreducibly *social* science, is thus the particular form assumed, in relation to the Subject, by the myth of *scientism*.

Myths, Lévi-Strauss has said, are machines for the suppression of time. They are also machines for its fabrication, as we could see in the examination of the role of the 'phylocultural complex'—an epic of scientistic myth-making—in the formation of developmental psychology. In our time, the myth of scientism manifests itself as a bureaucratic and alienated regime of Truth as power, and Science as the manipulation of symbolic values divested of social and historical reference. This regime shapes and governs our world, as well as representing its dominant order; detached from history and society, scientistic 'rationality' becomes a *weapon* for the suppression and fabrication of history, and for the superinscription of the values of the ruling order upon every arena of social reality.

The Representation myth has come, in the last decades, increasingly under attack within philosophy, the social sciences and literary theory. An important part of this challenge—or resistance—to the myth and regime of Representation takes its inspiration from the modernist insistence on the freedom of the sign to *escape* both the confines of Representation, and the mechanistic determinations of 'deep' form. Critical modernism's force derives not only from its negation of Representation, but also from its discovery of an eschatology of surfaces.[19]

Denial and negation, however, are not enough: what is repressed has a way of returning, unsolicited, in the guise of a new myth barely concealing its identity with the old one. The Anti-representation philosophies of the post-structuralists, arguing the dissolution of the signified into the signifier, seem (as Bickhard, 1987, also suggests in relation to pure hermeneutic philosophies) to lead to a kind of collectivist solipsism which is hardly more satisfactory than its individualist counterpart. The alternative, which I have tried in this book to pursue, is to *reconstruct* as well as to deconstruct representation; to understand it, that is, as pragmatically, semiotically and discursively *located* in social relations and practices. I cannot do better, then, to end with the following quotation[20] from Edward Said's book *Orientalism* (Said, 1985: 272–273):

> The real issue is whether indeed there can be a true representation of anything, or whether any and all representations, because they *are* representations, are embedded first in the language and then in the culture, institutions and political ambience of the representer. If the latter alternative is the correct one ... then we must be prepared to accept the fact that a representation is *eo ipso* implicated, intertwined, embedded, interwoven with a great many other things beside the "truth", which is itself a representation. What this must lead us to methodologically is to view representations (or mis-representations—the distinction is at best a matter of degree) as inhabiting a common field of play defined for them, not by some inherent subject matter alone, but by some common history, tradition, universe of discourse.

Notes

1. *Tao Te Ching*, transl. Chu'u Ta-Kao, London, George Allen and Unwin, 1970: 23.
2. *Moby Dick*, ch. 55 'Of the Monstrous Pictures of Whales'. Harmondsworth, Penguin, 1972: 371.
3. In this chapter, I employ the terms 'background' and 'background knowledge' interchangeably, in a sense equivalent to 'presuppositional background'. For a discussion of the epistemological issues involved here, see ch. 2.

4. I shall not directly address, therefore, the arguments regarding the 'partitioning' of event representations put forward by Nelson, 1983, 1985.
5. See ch. 1, note 8.
6. In common with some other development metatheories, the socio-naturalistic approach advocated in this book could be seen, in Pepper's terms, as an attempt to reconcile organicism with contextualism: though, as indicated in ch. 4, this does not mean that issues of mechanism are unimportant for general theoretical formulations. See Lerner and Kaufmann, 1985, 1986; Kendler, 1986.
7. For discussions of the complexities of this issue, see Morton, 1980; Hookway and Pettit, 1978.
8. Synecdoche—as instanced by the expression 'hands' to refer to factory workers—is a typically metonymic device. See also ch. 2.
9. The Linnaean order, too, is based upon difference and opposition; but within a different universe of discourse realized as a specific semantic field (taxonomy).
10. Following Jakobson's remarks above, it should be pointed out that the featural combination characterizing a prototype is concurrent, rather than chained or concatenated, but no less a metonymic structure for that fact.
11. 'Virtual' representations may be thought of as 'implicit' structures at the level *below* background proper.
12. See Paprotté and Sinha, 1987, for a more extended discussion of the sources and distribution of information utilized in discourse.
13. The implication of this is that prototypic representations underlying perceptual judgements, and the early stages of language acquisition, are developmental achievements rooted both in Piagetian sensori-motor development, and in the intersubjective completion of the conditions on representation (see ch. 2). This proposal is necessarily extremely condensed. If I had the space to do so, I would argue that this achievement—the first 'level' of human representational capacities—normally occurs in human infants at around nine months of age; and that it marks the earliest point at which, to use the terms of Wertsch's (1985) treatment of Vygotsky's theory, the 'natural' and the 'social' lines 'cross' in ontogenetic development.
14. See Reynolds, 1982, for an illuminating account of the evolution of primate constructional capacities.
15. The experiments reported in this chapter were designed and carried out jointly with Dr Norman Freeman, at the University of Bristol, as part of a larger study of the role of canonical representations, and their contextual variability, in cognitive and linguistic development (Freeman *et al.*, 1980; Freeman *et al.*, 1981; Freeman *et al.*, 1982; Lloyd *et al.*, 1981; Sinha and Freeman, 1981; Sinha, 1982a, 1983).
16. See also Walkerdine, 1982, for a critique of 'externalist' views of context which assume that it can be 'welded on' to otherwise unmodified Piagetian or other cognitivist modes of explanation.
17. The interpretations offered here are consistent with the account offered by Nelson, 1985, who emphasizes that 'both function and form are essential to the object concept; thus the either-or question is essentially meaningless' (p. 142). Nelson's 'functional core concept' theory, as well as the closely related

theory of canonicality proposed here, share many affinities with Duncker's theory of 'functional fixity', which was developed in the context of Gestalt psychology in the 1930s. Duncker, an assistant of Köhler, was expelled (against Köhler's public protests) from his German university post, on the pretext of 'communist activities', when the National Socialists came to power in 1933; he did not survive the war. This historical connection was pointed out to me independently by Steven Condliffe and Pim Levelt. See Ash, 1985; Duncker, 1945; Kessel and Bevan, 1985; Maier, 1930, 1931.

18. Other experiments, e.g. Freeman *et al.* (1980), have shown that possession of canonical 'proto-concepts' may be evidenced by infants as young as nine months of age.
19. See 'Language Writing' by Jerome McGann (*London Review of Books*, 15 November 1987).
20. Thanks to Nick Rowling for pointing me to this one.

Bibliography

Abelson, R. P. (1981). The psychological status of the script concept. *American Psychologist* 36: 715–729.

Althusser, L. (1971). *Lenin and Philosophy and Other Essays*. London, New Left Books.

Apel, K.-O. (1980). *Towards a Transformation of Philosophy*. London, Routledge & Kegan Paul.

Apostel, L., R. Pinkster, F. van Damme *et al.* (1987). *The Philosophy of Louis Apostel* (2 vols). Den Haag, Martinus Nijhoff.

Ash, M. (1985). Gestalt psychology: origins in Germany and reception in the United States. In C. Buxton (ed.) *Points of View in the Modern History of Psychology*. Orlando, Academic Press.

Atkinson, M. (1982). *Explanations in the Study of Child Language Development*. Cambridge, Cambridge University Press.

Bacon, F. (1960). *The New Organon and Related Writings*. New York, Liberal Arts Press.

Bailey, C. J. and R. Harris (1985). *Developmental Mechanisms of Language*. London, Pergamon.

Baldwin, J. (1897). *Le Développement Mentale Chez l'Enfant et Dans La Race*. Cited in Piaget (1979: 22).

Baldwin, J. (1902). *Development and Evolution*. London, Macmillan.

Barrett, M. (1982). Distinguishing between prototypes: the early acquisition of the meanings of object names. In S. A. Kuczaj (ed.) *Language Development: Syntax and Semantics*. Hillsdale, N.J., Lawrence Earlbaum.

Barthes, R. (1957). *Mythologies*. Paris, Editions Seuil.

Barthes, R. (1973). *Mythologies*. St Albans, Paladin.

Barthes, R. (1983). *Barthes: Selected Writings*. Oxford, Fontana/Collins.

Barthes, R. (1984). *Writing Degree Zero & Elements of Semiology*. London, Jonathan Cape.

Bartlett, Sir F. (1932). *Remembering*. Cambridge, Cambridge University Press.

Bates, E. and B. MacWhinney (1987). Competition, variation and language learning. In B. MacWhinney (ed.) *Mechanisms of Language Acquisition*. Hillsdale, N.J., Lawrence Earlbaum.

Bates, E., with L. Benigni, I. Bretherton, L. Camaioni, V. Volterra *et al.* (1979). *The Emergence of Symbols: cognition and communication in infancy*. New York, Academic Press.

Bateson, G. (1973). *Steps to an Ecology of Mind*. London, Paladin.

Beaugrande, R. de (1983). Freudian psychoanalysis and information processing: towards a new synthesis. Technical Report NL-22, University of Florida, Gainesville.

Beaugrande, R. de, and W. Dressler (1981). *Introduction to Text Linguistics*. London, Longmans.

Berger, J. (1980). *About Looking*. London, Writers and Readers.

Bernheimer, R. (1961). *The Nature of Representation: a phenomenological inquiry*. New York, New York University Press.

Bernstein, N. A. (1967). *The Co-ordination and Regulation of Movements*. Oxford, Pergamon Press.

Beveridge, M. (ed.) (1982). *Children Thinking through Language*. London, Edward Arnold.

Bhaskar, R. (1979). *The Possibility of Naturalism*. Brighton, Harvester Press.

Bickhard, M. (1987). The social nature of the functional nature of language. In M. Hickmann (ed.) *Social and Functional Approaches to Language and Thought*. Orlando, Academic Press.

Bierwisch, M. (1987). Cognitive linguistics: framework and topics. Paper presented at Max-Planck Institute for Psycholinguistics, October.

Bloom, L., K. Lifter and J. Broughton (1985). The convergence of early cognition and language in the second year of life: problems in conceptualization and measurement. In M. Barrett (ed.) *Children's Single-Word Speech*. Chichester, Wiley.

Bobrow, D. and D. Norman (1975). Some principles of memory schemata. In D. Bobrow and A. Collins (ed.) *Representation and Understanding*. New York, Academic Press.

Boden, M. (1977). *Artificial Intelligence and Natural Man*. Hassocks, Harvester Press.

Boden, M. (1979). *Piaget*. Glasgow, Fontana.

Boden, M. (1981). *Minds and Mechanisms: philosophical psychology and computational models*. Brighton, Harvester Press.

Bohm, D. (1980). *Wholeness and the Implicate Order*. London, Routledge & Kegan Paul.

Bowerman, M. (1978). The acquisition of word meaning: an investigation into some current conflicts. In N. Waterson and C. Snow (ed.) *The Development of Communication*. London, Wiley.

Bowerman, M. (1982). Reorganizational processes in lexical and syntactic development. In E. Wanner and L. Gleitmann (ed.) *Language Acquisition: the state of the art*. Cambridge, Cambridge University Press.

Bronfenbrenner, U. (1979). *The Ecology of Human Development*. London, Harvard University Press.

Brown, R. (1958). How shall a thing be called? *Psychological Review* 65: 14–21.

Brown, R. (1973). *A First Language: the early stages*. London, Allen & Unwin.

Bruner, J. (1971). The growth and structure of skill. In K. Connolly (ed.) *Mechanisms of Motor Skill Development*. London, Academic Press.

Bruner, J. (1974). *Beyond the Information Given: studies in the psychology of knowing*. London, Allen & Unwin.

Bruner, J. (1975). From communication to language: a psychological perspective. *Cognition* 3: 225–287.

Bruner, J. (1983). *Child's Talk*. Oxford, Oxford University Press.

Butterworth, G. (ed.) (1981). *Infancy and Epistemology: an evaluation of Piaget's theory*. Brighton, Harvester Press.

Butterworth, G. and P. Light (eds) (1982). *Social Cognition: studies in the development of understanding*. Brighton, Harvester Press.

Camaioni, L., C. de Lemos *et al.* (1985). *Questions on Social Explanation: Piagetian Themes Reconsidered*. Amsterdam, John Benjamins.

Carnap, R. (1956). *Meaning and Necessity*. Chicago, University of Chicago Press.

Castelfranchi, C. and I. Poggi (1987). Communication: beyond the cognitive approach and speech act theory. In J. Verschueren and M. Bertuccelli-Papi (ed.) *The Pragmatic Perspective*. Amsterdam, John Benjamins.

Changeux, J.-P. (1985). *Neuronal Man: the biology of mind*. Oxford, Oxford University Press.

Chomsky, N. (1957). *Syntactic Structures*. The Hague, Mouton.

Chomsky, N. (1959). Review of B. F. Skinner's 'Verbal Behavior'. *Language* 35: 265–8.

Chomsky, N. (1968). *Language and Mind*. New York, Harcourt, Brace & World.

Chomsky, N. (1980). Rules and Representations. *Behavioural and Brain Sciences* 3: 1–15.

Chomsky, N. (1981). *Lectures on Government and Binding*. Dordrecht, Foris.

Clark, E. (1973). Non-linguistic strategies and the acquisition of word meanings. *Cognition* 2: 161–182.

Clark, H. and S. Haviland (1977). Comprehension and the Given-New Contract. In R. Freedle (ed.) *Discourse Processes*. New Jersey, Ablex.

Clark, K. and M. Holquist (1984). *Mikhail Bakhtin*. Cambridge, Mass., Harvard University Press.

Cole, M., J. Gay, J. Glick and D. Sharp (1971). *The Cultural Context of Learning and Thinking: an exploration in experimental anthropology*. London, Methuen.

Cole, M., L. Hood and R. McDermott (1978). Ecological niche picking: ecological invalidity as an axiom of experimental psychology. New York, Laboratory of Comparative Human Cognition, Rockefeller University.

Costall, A. (1981). On how so much information controls so much behaviour: James Gibson's theory of direct perception. In G. Butterworth (ed.) *Infancy and Epistemology*. Brighton, Harvester Press.

Cranach, M. von and R. Harré (eds) (1982). *The Analysis of Action*. Cambridge, Cambridge University Press.

Cruse, D. (1986). *Lexical Semantics*. Cambridge, Cambridge University Press.

Cutting, J. (1982). Two ecological perspectives: Gibson versus Shaw and Turvey. *American Journal of Psychology* 95: 199–222.

Darwin, C. (1871). *The Descent of Man, and Selection in Relation to Sex*. London, John Murray.

Darwin, C. (1877). A biographical sketch of an infant. *Mind* 2: 285–294.

Dennett, D. (1979). *Brainstorms: Philosophical Essays on the Foundations of Cognitive Science*. Cambridge, Mass., MIT Press.

Derrida, J. (1982). *Margins of Philosophy*. Brighton, Harvester Press.

Dik, S. (1981). *Functional Grammar*. Dordrecht, Foris.

Dirven, R. and V. Fried (eds) (1987). *Functionalism in Linguistics. Linguistic and Literary Studies in Eastern Europe* (Vol. 20). Amsterdam, John Benjamins.

Donaldson, M. (1978). *Children's Minds*. Glasgow, Fontana.

Donnellan, K. (1966). Reference and definite descriptions. *Philosophical Review* 75: 281–304.

Dreyfus, H. (1972). *What Computers Can't Do: a critique of artificial reason*. New York, Harper & Row.

Dummett, M. (1973). *Frege: Philosophy of Language*. London, Duckworth.

Duncker, K. (1945). On problem solving. *Psychological Monographs* 58 (Whole No. 270).

Eco, U. (1976). *A Theory of Semiotics*. London, Macmillan.

Eco, U. (1979). *The Role of the Reader: explorations in the semiotics of texts*. London, Hutchinson.

Eco, U. (1983). *The Name of the Rose*. London, Secker & Warburg.

Eco, U. (1984). *Semiotics and the Philosophy of Language*. London, Macmillan.

Edelman, G. (1981). Group selection as the basis for higher brain function. In F. O. Schmitt *et al.* (ed.) *The Organization of the Cerebral Cortex*. Cambridge, Mass., MIT Press.

Ehrenreich, B. and D. English (1978). *For Her Own Good: 150 years of*

the experts' advice to women. New York, Anchor/Doubleday.
Erlich, V. (1969). *Russian Formalism: History–Doctrine.* The Hague, Mouton.
Fodor, J. (1972). Some reflections on L. S. Vygotsky's 'Thought and Language'. *Cognition* 1: 83–96.
Fodor, J. (1976). *The Language of Thought.* Hassocks, Harvester Press.
Fodor, J. (1980). Methodological solipsism considered as a research strategy in cognitive science. *The Behavioral and Brain Sciences* 3: 63–73.
Fodor, J. (1981). *Representations: philosophical essays on the foundations of cognitive science.* Brighton, Harvester Press.
Fodor, J. (1983). *The Modularity of Mind.* Cambridge, Mass., MIT Press.
Fodor, J. and Z. Pylyshyn (1981). How direct is visual perception? Some remarks on Gibson's 'Ecological approach'. *Cognition* 9: 139–96.
Fogelin, R. (1976). *Wittgenstein.* London, Routledge & Kegan Paul.
Foucault, M. (1970). *The Order of Things.* London, Tavistock Publications.
Freeman, N. (1980). *Strategies of Representation in Young Children: analysis of spatial skills and drawing processes.* London, Academic Press.
Freeman, N., C. Sinha and J. Stedmon (1982). All the cars—which cars? From word meaning to discourse analysis. In M. Beveridge (ed.) *Children Thinking Through Language.* London, Edward Arnold.
Freeman, N., C. Sinha and S. Condliffe (1981). Collaboration and confrontation with young children in language comprehension testing. In W. P. Robinson (ed.) *Communication in Development.* London, Academic Press.
Freeman, N., S. Lloyd and C. Sinha (1980). Infant search tasks reveal early concepts of containment and canonical usage of objects. *Cognition* 8: 243–262.
Frege, G. (1892). Über Sinn und Bedeutung. *Zeitschrift für Philosophie und Philosophische Kritik* 100: 25–50.
Freud, S. (1901). *The Psychopathology of Everyday Life.* London, Hogarth Press & Inst. Psychoanalysis.
Freud, S. (1922). *Introductory Lectures on Psychoanalysis.* London, Allen & Unwin & Int. Psychoanalytic Inst.
Freud, S. (1953–74). *The Standard Edition of the Complete Psychological Works of Sigmund Freud.* London, Hogarth Press & Inst. Psychoanalysis.
Freud, S. (1954). *The Interpretation of Dreams.* London, Allen & Unwin.
Furth, H. (1969). *Piaget and Knowledge.* Englewood Cliffs, N.J., Prentice-Hall.
Gardner, H. (1987). *The Mind's New Science: a history of the cognitive*

revolution. 2nd ed. New York, Basic Books.
Geert, P. van (1983). *The Development of Perception, Cognition and Language.* London, Routledge & Kegan Paul.
Geertz, C. (1973). *The Interpretation of Cultures.* New York, Basic Books.
Gelder, B. de (1985). The cognitivist conjuring trick or how development vanished. In C.-J. Bailey and R. Harris (ed.) *Developmental Mechanisms of Language.* Oxford, Pergamon Press.
Gelder, B. de (1987). Wat is er cognitief aan informatieverweking? [What is cognitive about information processing?]. *Psychologie en Maatschappij* 40: 259–271.
Gibson, J. (1979). *The Ecological Approach to Visual Perception.* Boston, Houghton Mifflin.
Giles, H. (1973). Accent mobility: a model and some data. *Anthropological Linguistics* 15: 87–105.
Givón, T. (ed.) (1979). *Discourse and Syntax: syntax and semantics.* New York, Academic Press.
Goethe, J. von (1970). *Theory of Colours.* Cambridge, Mass., MIT Press.
Gombrich, E. (1985). Scenes in a Golden Age: masters of seventeenth-century Dutch genre painting. *New York Review* 32 10: 20–22.
Gould, S. (1977). *Ontogeny and Phylogeny.* Cambridge, Mass., Harvard University Press.
Gould, S. (1987). *Time's Arrow, Time's Cycle: myth and metaphor in the discovery of geological time.* London, Harvard University Press.
Greenfield, P. (1978). Informativeness, presupposition and semantic choice in single-word utterances. In N. Waterson and C. Snow (ed.) *The Development of Communication.* Chichester, Wiley.
Greenfield, P. (1982). The role of perceived variability in the transition to language. *Journal of Child Language* 9: 1–12.
Greenspan, S. (1979). *Intelligence and Adaptation: an integration of Psychoanalytic and Piagetian developmental psychology.* New York, International Universities Press.
Grice, H. (1975). Logic and conversation. In P. Cole and J. Morgan (ed.) *Syntax and Semantics 3: Speech Acts.* New York, Academic Press.
Habermas, J. (1970). On systematically distorted communication. *Inquiry* 13: 205–218.
Habermas, J. (1971). *Towards a Rational Society.* London, Heinemann.
Haeckel, E. (1874). *The Evolution of Man: a popular exposition of the principal points of human ontogeny and phylogeny.* New York, International Science Library.
Hamlyn, D. (1978). *Experience and the Growth of Understanding.* London, Routledge & Kegan Paul.
Hamlyn, D. (1982). What exactly is social about the origins of understanding? In G. Butterworth and P. Light (ed.) *Social Cognition:*

studies of the development of understanding. Brighton, Harvester Press.

Harland, R. (1987). *Superstructuralism: the philosophy of structuralism and post-structuralism*. London, Methuen.

Haugeland, J. (1978). The nature and plausibility of cognitivism. *The Behavioral and Brain Sciences* 1: 215–226.

Henriques, J. and C. Sinha (1974). Language and revolution: Volosinov's 'Marxism and the Philosophy of Language'. *Ideology and Consciousness* 1: 93–96.

Hickmann, M. (ed.) (1987). *Social and Functional Approaches to Language and Thought*. New York, Academic Press.

Ho, M.-W. (1984). Environment and heredity in development and evolution. In M.-W. Ho and P. Saunders (ed.) *Beyond Neo-Darwinism: an introduction to the new evolutionary paradigm*. London, Academic Press.

Ho, M.-W. and P. Saunders (eds) (1984). *Beyond Neo-Darwinism: an introduction to the new evolutionary paradigm*. London, Academic Press.

Hofstadter, D. (1979). *Gödel, Escher, Bach: An Eternal Golden Braid*. Brighton, Harvester Press.

Holzman, L. H. (1985). Pragmatism and dialectical materialism in language development. In K. Nelson (ed.) *Children's Language* (Vol. 5). Hillsdale, N.J., Lawrence Earlbaum Associates.

Hookway, C. and P. Pettit (eds.) (1978). *Action and Interpretation: studies in the philosophy of the social sciences*. Cambridge, Cambridge University Press.

Hutt, C. (1966). Exploration and play in children. *Symposia of the Zoological Society of London* 18: 61.

Ingleby, D. (1983). Freud and Piaget: the phoney war. *New Ideas in Psychology* 1: 123–144.

Irigaray, L. (1978). Women's exile. *Ideology and Consciousness* 1: 57–76.

Jakobson, R. (1956). Two aspects of language and two types of aphasic disturbance. In R. Jakobson and M. Halle (ed.) *Fundamentals of Language*. The Hague, Mouton.

Jakobson, R. and J. Tynjanov (1985). Problems in the study of language and literature. In K. Pomorska and S. Rudy (ed.) *Roman Jakobson: Verbal Art, Verbal Sign, Verbal Time*. Oxford, Basil Blackwell.

Jerne, N. (1967). Antibodies and learning: selection versus instruction. In G. Quarton, T. Melchnuk and F. O. Schmitt (ed.) *The neurosciences: a study program*. New York, Rockefeller University Press.

John-Steiner, V. (1987). *Notebooks of the Mind: explorations of thinking*. New York, Harper & Row.

Johnson-Laird, P. (1983). *Mental Models*. Cambridge, Cambridge

University Press.

Johnson-Laird, P. and A. Garnham (1980). Descriptions and discourse models. *Linguistics and Philosophy* 3: 371–393.

Kant, I. (1929). *Critique of Pure Reason.* London, Macmillan.

Karmiloff-Smith, A. (1979). *A Functional Approach to Child Language: a study of determiners and reference.* Cambridge, Cambridge University Press.

Karmiloff-Smith, A. (1987). Function and process in comparing language and cognition. In M. Hickmann (ed.) *Social and Functional Approaches to Language and Thought.* New York, Academic Press.

Karpatschof, B. (1982). Artificial intelligence or artificial signification? *Journal of Pragmatics* 6: 293–304.

Katz, J. and J. Fodor (1963). The structure of a semantic theory. *Language* 39: 170–210.

Kendler, T. (1986). World views and the concept of development: a reply to Lerner and Kauffman. *Developmental Review* 6: 80–95.

Kessel, F. and W. Bevan (1985). Notes toward a history of cognitive psychology. In C. Buxton (ed.) *Points of View in the Modern History of Psychology.* Orlando, Academic Press.

Kripke, S. (1980). *Naming and Necessity.* Oxford, Basil Blackwell.

Kristeva, J. (1970). *La Texte du Roman.* The Hague, Mouton.

Kristeva, J. (1986). *The Kristeva Reader.* Oxford, Basil Blackwell.

Kuczaj, S. (ed) (1984). *Discourse Development.* New York, Springer-Verlag.

Kuhn, T. (1970). *The Structure of Scientific Revolutions.* Chicago, University of Chicago Press.

Lacan, J. (1966). *Ecrits.* Paris, Editions du Seuil.

Lane, H. (1977). *The Wild Boy of Aveyron.* London, Allen & Unwin.

Langacker, R. (1987). *Foundations of Cognitive Grammar* (Vol. 1) *Theoretical Prerequisites.* Stanford, Stanford University Press.

Larrain, J. (1979). *The Concept of Ideology.* London, Hutchinson.

Larsen, S. F. (1985). Specific background knowledge and knowledge updating. In J. Allwood and E. Hjelmquist (ed.) *Foregrounding Background.* Lund, Doxa.

Lemos, C. de (1981). Interactional processes in the child's construction of language. In W. Deutsch (ed.) *The Child's Construction of Language.* London, Academic Press.

Lenin, V. I. (1961). *Philosophical Notebooks* (Collected Works, Vol. 38). London, Lawrence & Wishart.

Leontiev, A. (1981). *Problems of the Development of the Mind.* Moscow, Progress Publishers.

Lerner, R. and M. Kaufmann (1985). The concept of development in contextualism. *Developmental Review* 5: 309–333.

Lerner, R. and M. Kaufmann (1986). On the metatheoretical relativism of analyses of metatheoretical analyses: a critique of Kendler's comments. *Developmental Review* 6: 96–106.

Lévi-Strauss, C. (1969). *The Elementary Structures of Kinship*. Boston, Beacon Press.

Levinson, S. (1983). *Pragmatics*. Cambridge, Cambridge University Press.

Lévy-Bruhl, L. (1918). *Les Fonctions mentales dans les sociétés inférieures*. Paris, Alcan.

Lewin, K. (1946). Behaviour and development as a function of the total situation. In L. Carmichael (ed.) *Manual of Child Psychology*. New York, Wiley.

Lloyd, S., C. Sinha and N. Freeman (1981). Spatial reference systems and the canonicality effect in infant search. *Journal of Experimental Child Psychology* 32: 1–10.

Lock, A. (ed.) (1978). *Action, Gesture and Symbol: the emergence of language*. London, Academic Press.

Lock, A. (1980). *The Guided Reinvention of Language*. London, Academic Press.

Lovejoy, A. (1936). *The Great Chain of Being*. Cambridge, Mass., Harvard University Press.

Luria, A. (1973). *The Working Brain: an introduction to neurospsychology*. Harmondsworth, Penguin.

Lyons, J. (1977). *Semantics*. Cambridge, Cambridge University Press.

McCulloch, W. and W. Pitts (1943). A logical calculus of the ideas immanent in nervous activity. *Bulletin of Mathematical Biophysics* 5: 115–133.

MacLean, P. (1972). Cerebral evolution and emotional processes. *Annals of the New York Academy of Sciences* 193: 137–149.

Macnamara, J. (1982). *Names for Things: a study of human learning*. Cambridge, Mass., MIT Press.

McNeill, D. (1979). *The Conceptual Basis of Language*. Hillsdale, N.J., Lawrence Earlbaum.

Maier, N. (1930). Reasoning in humans. I. On direction. *Journal of Comparative Psychology* 10: 115–143.

Maier, N. (1931). Reasoning in humans. II. The solution of a problem and its appearance in consciousness. *Journal of Comparative Psychology* 12: 181–194.

Malinowski, B. (1930). The problem of meaning in primitive languages. In C. Ogden and I. A. Richards (ed.) *The Meaning of Meaning* (2nd edition). London, Routledge & Kegan Paul.

Mandler, J. (1984). Representation. In P. Mussen, J. Flavell and E. Markman (ed.) *Manual of Child Psychology*, (Vol. 2) *Cognitive Development*. New York, Wiley.

Markman, E. (1976). Children's difficulty with word-referent differentiation. *Child Development* 47: 742–749.

Marková, I. (1982). *Paradigms, Thought and Language*. Chichester, Wiley.

Marr, D. (1982). *Vision: a computational investigation into the human representation and processing of visual information*. San Francisco, Freeman.

Marshall, J. (1980). The new organology. *Behavioural and Brain Sciences* 3: 23–25.

Marx, K. (1975). *Early Writings*. Harmondsworth and London, Penguin & New Left Review.

Marx, K. (1976). *Capital* (Vol. 1). Harmondsworth, Penguin.

Mead, G. (1934). *Mind, Self and Society*. Chicago, University of Chicago Press.

Mehler, J. (1972). Knowing by unlearning. Paper presented at the International Seminar of the Centre International d'Etudes de Bio-Anthropologie et d'Anthropologie Fondamentale.

Merquior, J. (1986). *From Prague to Paris: a critique of structuralist and post structuralist thought*. London, Verso.

Miller, M. (1987). Argumentation and cognition. In M. Hickmann (ed.) *Social and Functional Approaches to Language and Thought*. Orlando, Academic Press.

Minsky, M. (1975). A framework for representing knowledge. In P. Winston (ed.) *The Psychology of Computer Vision*. New York, McGraw-Hill.

Moore, T. and C. Carling (1982). *Understanding Language: towards a post-Chomskyan linguistics*. London, Macmillan.

Morris, C. (1938). Foundations of the theory of signs. In O. Neurath, R. Carnap and C. Morris (ed.) *International Encyclopaedia of Unified Science*. Chicago, University of Chicago Press.

Morton, A. (1980). *Frames of Mind: constraints on the common sense conception of the mental*. Oxford, Clarendon Press.

Neisser, U. (1976). *Cognition and Reality*. San Francisco, Freeman.

Nelson, K. (1974). Concept, word and sentence: inter-relations in acquisition and development. *Psychological Review* 91: 267–284.

Nelson, K. (1983). The conceptual basis for language. In T. Seiler and W. Wannenmacher (ed.) *Concept Development and the Development of Word Meaning*. Berlin, Springer-Verlag.

Nelson, K. (1985). *Making Sense: the acquisition of shared meaning*. Orlando, Academic Press.

Nelson, K. E. and K. Nelson (1978). Cognitive pendulum swings and their linguistic realization. In K. E. Nelson (ed.) *Children's Language* 1. New York, Gardner Press.

Nunberg, G. (1979). The non-uniqueness of semantic solutions: polysemy. *Linguistics and Philosophy* 3: 143–184.
Nuyts, J. (1987). Pragmatics and Cognition: on explaining language. In J. Verschueren and M. Bertuccelli-Papi (ed.) *The Pragmatic Perspective*. Amsterdam, John Benjamins.
Oatley, K. (1978). *Perceptions and Representations: the theoretical bases of brain research and psychology*. London, Methuen.
Olson, D. (1977). From utterance to text: the bias of language in speech and writing. *Harvard Educational Review* 47: 257–281.
Palermo, D. (1982). Theoretical issues in semantic development. In S. Kuczaj (ed.) *Language Development: Syntax and Semantics*. Hillsdale, N.J., Lawrence Earlbaum.
Paprotté, W. and C. Sinha (1987). Functional sentence perspective in discourse and language acquisition. In R. Dirven and V. Fried (ed.) *Functionalism in Linguistics*. Amsterdam, John Benjamins.
Parret, H. (1983). *Semiotics and Pragmatics*. Amsterdam, John Benjamins.
Parret, H. (1985). Time, space and actors: the pragmatics of development. In C.-J. Bailey and R. Harris (ed.) *Developmental Mechanisms of Language*. Oxford, Pergamon Press.
Pepper, S. (1942). *World Hypotheses*. Berkeley, University of California Press.
Petöfi, J. (1987). Models in descriptive meaning interpretation. In J. Verschueren and M. Bertuccelli-Papi (ed.) *The Pragmatic Perspective*. Amsterdam, John Benjamins.
Piaget, J. (1953). *The Origins of Intelligence in the Child*. London, Routledge & Kegan Paul.
Piaget, J. (1959). *The Language and Thought of the Child*. London, Routledge & Kegan Paul.
Piaget, J. (1962). *Play, Dreams and Imitation*. London, Routledge & Kegan Paul.
Piaget, J. (1971). *Structuralism*. London, Routledge & Kegan Paul.
Piaget, J. (1977a). *The Essential Piaget: an interpretive reference and guide. Edited by H. Gruber and J. J. Vonèche*. New York, Basic Books.
Piaget, J. (1977b). *The Development of Thought: equilibration of cognitive structures*. Oxford, Basil Blackwell.
Piaget, J. (1979). *Behaviour and Evolution*. London, Routledge & Kegan Paul.
Piaget, J. and B. Inhelder (1973). *Memory and Intelligence*. London, Routledge & Kegan Paul.
Piattelli-Palmarini, M. (ed.) (1980). *Language and Learning*. London, Routledge & Kegan Paul.
Pinker, S. (1979). Formal models of language learning. *Cognition* 7:

217–283.

Plunkett, K. and S. F. Larsen (eds) (1988). *Computers, Cognition and Context*. Hillsdale, N.J., Lawrence Earlbaum.

Pribram, K. and M. Gill (1976). *Freud's 'Project' Reassessed*. London, Hutchinson.

Prigogine, I. and I. Stengers (1984). *Order Out of Chaos: man's new dialogue with nature*. London, Heinemann.

Putnam, H. (1975). *Mind, Language, and Reality (Philosophical Papers*, (Vol. 1). Cambridge, Cambridge University Press.

Putnam, H. (1981). *Reason, Truth and History*. Cambridge, Cambridge University Press.

Reynolds, P. C. (1981). *On the Evolution of Human Behaviour: the argument from animals to man*. Berkeley, University of California Press.

Reynolds, P. (1982). The primate constructional system: the theory and description of instrumental object use in humans and chimpanzees. In M. von Cranach and R. Harré (ed.) *The Analysis of Action*. Cambridge, Cambridge University Press.

Richards, M. P. M. (ed.) (1974). *The Integration of the Child into a Social World*. Cambridge, Cambridge University Press.

Richards, M. and P. Light (eds) (1986). *Children of Social Worlds*. London, Polity.

Richardson, K. and C. Sinha (1987). Course E362 Text Language and Cognitive Development in the School Years. Milton Keynes, The Open University Press.

Riley, D. (1978). Developmental psychology, biology and marxism. *Ideology and Consciousness* 4: 73–91.

Riley, D. (1983). *War in the Nursery*. London, Virago.

Robinson, P. (1987). Freud and the feminists. *Raritan* 6: 43–61.

Rommetveit, R. and R. Blakar (eds) (1979). *Studies of Language, Thought and Verbal Communication*. London, Academic Press.

Rosch E. (1977). Classification of real-world objects: origins and representations in cognition. In P. N. Johnson-Laird and P. C. Wason (ed.) *Thinking: Readings in Cognitive Science*. Cambridge, Cambridge University Press.

Rose, S. (ed.) (1982a). *Against Biological Determinism*. London, Allison & Busby.

Rose, S. (ed.) (1982b). *Towards a Liberatory Biology*. London, Allison & Busby.

Rose, S., R. Lewontin and L. Kamin (1984). *Not In Our Genes: biology, ideology and human nature*. Harmondsworth, Penguin.

Rosenfield, I. (1986). Neural Darwinism: a new approach to memory and perception. *New York Review* 9: October, 21–27.

Rotman, B. (1978). *Mathematics: an essay in semiotics*. Bristol, University of Bristol Monograph.
Rumelhart, D., J. McClelland and the PDP Research Group (1986). *Parallel Distributed Processing: explorations in the microstructure of cognition* (2 vols). Cambridge, Mass., MIT Press.
Russell, J. (1978). *The Acquisition of Knowledge*. London, Macmillan.
Sahlins, M. (1976). *The Use and Abuse of Biology: an anthropological critique of sociobiology*. Ann Arbor, University of Michigan Press.
Said, E. (1985). *Orientalism*. Harmondsworth, Penguin.
Saussure, F. de (1966). *Cours de Linguistique Générale*. New York, McGraw-Hill.
Schank, R. (1982). *Dynamic Memory*. Cambridge, Cambridge University Press.
Schank, R. and R. Abelson (1977). *Scripts, Plans, Goals and Understanding*. Hillsdale, N.J., Lawrence Earlbaum.
Schilcher, F. von and N. Tennant (1984). *Philosophy, Evolution and Human Nature*. London, Routledge & Kegan Paul.
Schmidt, R. (1975). A schema theory of discrete motor learning. *Psychological Review* **82**: 225–260.
Scribner, S. (1985). Vygotsky's uses of history. In J. Wertsch (ed.) *Culture, Communication, and Cognition: Vygotskian perspectives*. Cambridge, Cambridge University Press.
Scribner, S. and M. Cole (1978). Literacy without schooling: testing for intellectual effects. Vai Literacy Project Working Paper 2, New York, Laboratory of Comparative Human Cognition, Rockefeller University.
Searle, J. (1980). Minds, brains and programs. *Behavioral and Brain Sciences* **3**: 417–424.
Searle, J. (1986). Introductory essay: notes on conversation. In D. Ellis and W. Donohue (ed.) *Contemporary Issues in Language and Discourse Processes*. Hillsdale, N.J., Lawrence Earlbaum Associates.
Sechenov, I. (1935). *Selected Works*. Moscow-Leningrad, State Publishing House for Biological and Medical Literature.
Seltman, M. and P. Seltman (1985). *Piaget's Logic: a critique of genetic epistemology*. London, George Allen & Unwin.
Shaw, R. and M. Turvey (1980). Methodological realism. *Behavioral and Brain Sciences* **3**: 94–97.
Shaw, R., M. Turvey and W. Mace (1982). Ecological psychology: the consequences of a commitment to realism. In W. Weimer and D. Palermo (ed.) *Cognition and the Symbolic Processes* (Vol. 2). Hillsdale, N.J., Lawrence Earlbaum.
Shotter, J. (1984). *Social Accountability and Selfhood*. Oxford, Basil Blackwell.
Shotter, J. and J. Newson (1982). An ecological approach to cognitive

development: implicate orders, joint action and intentionality. In G. Butterworth and P. Light (ed.) *Social Cognition: studies in the development of understanding*. Brighton, Harvester Press.

Silverman, D. and B. Torode (1980). *The Material Word: some theories of language and its limits*. London, Routledge & Kegan Paul.

Sinha, C. (1982a). Representational development and the structure of action. In G. Butterworth and P. Light (ed.) *Social Cognition: studies in the development of understanding*. Brighton, Harvester Press.

Sinha, C. (1982b). Negotiating boundaries: psychology, biology and society. In S. Rose (ed.) *Towards a Liberatory Biology*. London, Allison & Busby.

Sinha, C. (1983). Background knowledge, presupposition and canonicality. In T. Seiler and W. Wannenmacher (ed.) *Concept Development and the Development of Word Meaning*. Berlin, Springer-Verlag.

Sinha, C. (1984). A socio-naturalistic approach to human development. In M-W. Ho and P. Saunders (ed.) *Beyond Neo-Darwinism: an introduction to the new evolutionary paradigm*. London, Academic Press.

Sinha, C. and B. Carabine (1981). Interactions between lexis and discourse in conservation and comprehension tasks. *Journal of Child Language* 8: 109–129.

Sinha, C. and N. Freeman (1981). L'effet canonique: une nouvelle perspective sur le concept objectal et sa représentation linguistique. *Bulletin de Psychologie* 24: 713–723.

Sinha, C. and V. Walkerdine (1975). Functional and perceptual aspects of the acquisition of spatial relational terms. University of Bristol, unpublished mimeo.

Sinha, C. and V. Walkerdine (1978). Children, logic and learning. *New Society* 43: 62–64.

Slobin, D. (1981). The origin of grammatical encoding of events. In W. Deutsch (ed.) *The Child's Construction of Language*. London, Academic Press.

Sontag, S. (1982). *A Susan Sontag Reader*. New York, Farrar, Strauss & Girar.

Sperber, D. and D. Wilson (1986). *Relevance: communication and cognition*. Oxford, Blackwell.

Steiner, P. (1984). *Russian Formalism: a metapoetics*. London, Cornell University Press.

Stich, S. (1978). Beliefs and subdoxastic states. *Philosophy of Science* 45: 499–518.

Sulloway, F. (1979). *Freud: Biologist of the Mind*. New York, Basic Books.

Thelen, E., J. Scott Kelso and A. Fogel (1987). Self-organizing systems and infant motor development. *Developmental Review* 7: 39–65.

Thompson, E. (1978). *The Poverty of Theory.* London, Merlin Press.
Thomson, J. (1920). *The System of Animate Nature.* London, William & Norgate.
Todorov, T. (1984). *Mikhail Bakhtin: the dialogical principle.* Manchester, Manchester University Press.
Trevarthen, C. and P. Hubley (1978). Secondary intersubjectivity: confidence, confiding and acts of meaning in the first year. In A. Lock (ed.) *Action, Gesture and Symbol: the emergence of language.* London, Academic Press.
Tulving, E. (1972). Episodic and semantic memory. In E. Tulving and W. Donaldson (ed.) *Organization and Memory.* New York, Academic Press.
Uexküll, J. von (1957). A stroll through the world of animals and men. In C. H. Schiller (ed.) *Instinctive Behaviour.* London, Methuen.
Urwin, C. (1985). Constructing motherhood: the persuasion of normal development. In C. Steedman, C. Urwin and V. Walkerdine (ed.) *Language, Gender and Childhood.* London, Routledge & Kegan Paul.
Venn, C. (1984). The subject of psychology. In Henriques, J., W. Hollway, C. Urwin, C. Venn and V. Walkerdine *Changing the Subject.* London, Methuen.
Venn, C. and V. Walkerdine (1978). The acquisition and production of knowledge: Piaget's theory reconsidered. *Ideology and Consciousness* 3: 67–94.
Volosinov, V. N. (1973). *Marxism and the Philosophy of Language.* New York, Seminar Press.
Vygotsky, L. S. (1978). *Mind in Society: the development of higher mental processes.* Cambridge, Mass., Harvard University Press.
Vygotsky, L. S. (1981). The Genesis of Higher Mental Functions. In J. Wertsch (ed.) *The Concept of Activity in Soviet Psychology.* N.Y., M. E. Sharpe, Inc.
Vygotsky, L. S. (1986). *Thought and Language.* Cambridge, Mass., MIT Press.
Waddington, C. H. (1975). *The Evolution of an Evolutionist.* Edinburgh, Edinburgh University Press.
Waddington, C. H. (1977). *Tools for Thought.* St Albans, Paladin.
Walkerdine, V. (1982). From context to text: a psychosemiotic approach to abstract thought. In M. Beveridge (ed.) *Children Thinking Through Language.* London, Edward Arnold.
Walkerdine, V. (1984). Developmental psychology and the child-centred pedagogy: the insertion of Piaget into early education. In Henriques, J., W. Hollway, C. Urwin, C. Venn and V. Walkerdine *Changing the Subject.* London, Methuen.
Walkerdine, V. (1985). On the regulation of speaking and silence:

subjectivity, class and gender in contemporary schooling. In C. Steedman, C. Urwin and V. Walkerdine (ed.) *Language, Gender and Childhood*. London, Routledge & Kegan Paul.

Walkerdine, V. (1988). *The Mastery of Reason*. London, Methuen.

Walkerdine, V. and G. Corran (1979). Cognitive development: a mathematical experience? Paper presented to the BPS Developmental Section Annual Conference, Southampton.

Walkerdine, V. and C. Sinha (1978). The internal triangle: language, reasoning and the social context. In I. Marková (ed.) *The Social Context of Language*. Chichester, Wiley.

Weismann, A. (1972). The continuity of the germ-plasm as the foundation of a theory of heredity. In J. A. Moore (ed.) *Readings in Heredity and Development*. Oxford, Oxford University Press.

Weizenbaum, J. (1976). *Computer Power and Human Reason*. San Francisco, Freeman.

Wells, G. (ed.) (1981). *Learning through Interaction: the study of language development*. Cambridge, Cambridge University Press.

Wertsch, J. (1985). *Vygotsky and the Social Formation of Mind*. Cambridge, Mass., Harvard University Press.

Wexler, K. (1982). A principle theory for language acquisition. In E. Wanner and L. Gleitman (ed.) *Language Acquisition: the state of the art*. Cambridge, Cambridge University Press.

White, S. (1983). The idea of development in developmental psychology. In R. Lerner (ed.) *Developmental Psychology: Historical and Philosophical Perspectives*. Hillsdale, N.J., Lawrence Earlbaum Associates.

Wilcox, S. and D. Palermo (1975). 'In', 'on' and 'under' revisited. *Cognition* 3: 245–254.

Wilden, A. (1972). *System and structure: Essays in Communication and Exchange*. London, Tavistock.

Williams, R. (1977). *Marxism and Literature*. Oxford, Oxford University Press.

Williams, R. (1986). The uses of cultural theory. *New Left Review* 158: 19–31.

Wilson, E. O. (1975). *Sociobiology: the new synthesis*. Cambridge, Mass., MIT Press.

Wittgenstein, L. (1953). *Philosophical Investigations*. Oxford, Blackwell.

Wittgenstein, L. (1961). *Tractatus Logico-Philosophicus*. London, Routledge & Kegan Paul.

Wolff, P. (1960). *The Developmental Psychologies of Jean Piaget and Psychoanalysis*. New York, International Universities Press.

Wood, D., J. Bruner and G. Ross (1976). The role of tutoring in problem solving. *Journal of Child Psychology and Psychiatry* 17: 89–100.

Yakubinsky, L. (1923). *O dialogicheskoi rechi* [*On dialogic speech*]. Petrograd, Trudy Foneticheskogo Instituta Prakticheskogo Izucheniya Yazykov.

Name Index

Subject Index